MATHEMATICAL ANALYSIS:
A Fundamental and Straightforward Approach

Mathematics and its Applications

Series Editor: G. M. BELL, Professor of Mathematics,
King's College London (KQC), University of London

Statistics and Operational Research

Editor: B. W. CONOLLY, Professor of Operational Research,
Queen Mary College, University of London

Mathematics and its applications are now awe-inspiring in their scope, variety and depth. Not only is there rapid growth in pure mathematics and its applications to the traditional fields of the physical sciences, engineering and statistics, but new fields of application are emerging in biology, ecology and social organisation. The user of mathematics must assimilate subtle new techniques and also learn to handle the great power of the computer efficiently and economically.

The need for clear, concise and authoritative texts is thus greater than ever and our series will endeavour to supply this need. It aims to be comprehensive and yet flexible. Works surveying recent research will introduce new areas and up-to-date mathematical methods. Undergraduate texts on established topics will stimulate student interest by including applications relevant at the present day. The series will also include selected volumes of lecture notes which will enable certain important topics to be presented earlier than would otherwise be possible.

In all these ways it is hoped to render a valuable service to those who learn, teach, develop and use mathematics.

Mathematics and its Applications

Series Editor: G. M. BELL, Professor of Mathematics, King's College London (KQC), University of London

Series continued at back of book

MATHEMATICAL ANALYSIS:
A Fundamental and Straightforward Approach

DAVID S. G. STIRLING, B.Sc., Ph.D.
Lecturer in Mathematics, University of Reading

ELLIS HORWOOD LIMITED
Publishers · Chichester

Halsted Press: a division of
JOHN WILEY & SONS
New York · Chichester · Brisbane · Toronto

First published in 1987 by
ELLIS HORWOOD LIMITED
Market Cross House, Cooper Street,
Chichester, West Sussex, PO19 1EB, England
The publisher's colophon is reproduced from James Gillison's drawing of the ancient Market Cross, Chichester.

Distributors:

Australia and New Zealand:
JACARANDA WILEY LIMITED
GPO Box 859, Brisbane, Queensland 4001, Australia
Canada:
JOHN WILEY & SONS CANADA LIMITED
22 Worcester Road, Rexdale, Ontario, Canada
Europe and Africa:
JOHN WILEY & SONS LIMITED
Baffins Lane, Chichester, West Sussex, England
North and South America and the rest of the world:
Halsted Press: a division of
JOHN WILEY & SONS
605 Third Avenue, New York, NY 10158, USA

© **1987 D. S. G. Stirling/Ellis Horwood Limited**

British Library Cataloguing in Publication Data
Stirling, David S. G.
Mathematical analysis. —
(Ellis Horwood series in mathematics and its applications)
1. Calculus
I. Title
515   QA303

Library of Congress Card No. 87–13500

ISBN 0–85312–996–7 (Ellis Horwood Limited — Library Edn.)
ISBN 0–7458–0262–1 (Ellis Horwood Limited — Student Edn.)
ISBN 0–470–20903–8 (Halsted Press)

Printed in Great Britain by Unwin Bros. of Woking

Table of Contents

Author's Preface

Analysis tackles the issues which were fudged in the development of the calculus. With the recent trend away from formal proof in school, it may not be evident to students beginning higher education that there is a problem to be attended to here. Indeed, some school leavers have seen virtually none of the ideas of proof and do not necessarily accept that it is a vital part of mathematics. This book was written in acknowledgement that most present-day students have this background.

The main aim of the book is to present the accepted core material of analysis in such a way that the development appears fairly natural to the reader. A detailed discussion of the real number system, which is necessarily technical, is postponed until other matters have highlighted the need for it, while I have tried to maximise the number of results whose value can be appreciated from a standpoint other than that of the analyst, so that the subject is not seen as merely self-serving. This approach, while it could not be sustained throughout a degree course, seems to be correct for the start of a subject. The technical jargon of analysis cannot sensibly be avoided but it can be minimised and I have taken the view that a definition is not worth the sacrifice of memory unless it is used often.

The principal difference between this book and many others is that attention is devoted not only to giving proofs but to indicating how one might construct these proofs, a rather different process from appreciating the final product. The completeness of the real number system is assumed in the form of Dedekind's axiom of continuity, because this is more plausible than some of its immediate consequences.

Logically, this book presumes no knowledge of calculus, but it would be rather pointless to start analysis without that, and I have tacitly relied on calculus for some of the motivation. This is particularly true of Chapter 11, on functions of several variables, where the experience of grappling with the problems which arise in practice is a necessary supplement to the theory.

The book contains many problems for the reader to solve, designed to illustrate the main points or to force attention onto the subtler ones. Tackling these problems is an essential part of reading the book although the starred problems may be regarded as optional, being more difficult or more peripheral than the others.

I should like to thank my colleagues at Reading, especially David White, for comments and useful conversations over the years and the students who have been subjected to this course for their comments. In particular, I am grateful to Michael Sewell for the final impetus which made me write it, to Robin Dixon for help with the diagrams, to Joyce Bird and Rosemary Pellew for deciphering my handwriting and typing it so expertly and to Ellis Horwood and his staff for their editorial and production cooperation.

Reading, February 1987   David Stirling

CHAPTER 1

The Need for Proof

Although mathematics is usually thought of as a science, it differs from most of science in one important respect—it is not based on empirical results which may later be altered by improved evidence. Thus physics, for example, is based on the prediction of various consequences of the basic 'laws' of the subject, but since these laws are derived from experiment and observation, they may be revised from time to time if their consequences turn out to conflict with what happens in the real world. Mathematics, on the other hand, is not based on experiment but is a more abstract creation whose results are true in a way that is not subject to later revision. A mathematical result, once established, is known to be true without reservation.

At its heart, mathematics is about numbers, which are already an abstraction: the thing that two sheep and two apples have in common (the 'two-ness') is abstract. Once we have accepted the idea of number, we have to find out the properties of the system we have created, that is, deduce them from our basic ideas. After we have discovered these, they remain true for all time and are not subject to the periodic revision that occurs with scientific laws. (Nevertheless, although the mathematics remains constant, it is at least conceivable that revisions in scientific laws could dramatically alter its usefulness.)

Having said that mathematics is a man-made structure in which results are deduced from some basic properties, let us consider the sort of processes involved. To fix our ideas here, let us look at a particular set of problems:

(i) $\sqrt{(x+3)} = \sqrt{(1-x)} + \sqrt{(1+x)}$,
(ii) $\sqrt{(x+3)} = \sqrt{(1-x)} - \sqrt{(1+x)}$,
(iii) $\sqrt{(x+3)} = \sqrt{(1+x)} - \sqrt{(1-x)}$.

Before starting, recall that the $\sqrt{y}$ sign denotes the non-negative square root of y.

Let us start with the first equation, $\sqrt{(x+3)} = \sqrt{(1-x)} + \sqrt{(1+x)}$. Squaring both sides gives

$$x + 3 = 1 - x + 2\sqrt{(1-x)} \cdot \sqrt{(1+x)} + (1+x),$$

and, on rearranging, we obtain

$$x + 1 = 2\sqrt{(1 - x)} \cdot \sqrt{(1 + x)}.$$

Squaring now yields

$$x^2 + 2x + 1 = 4 - 4x^2$$

whence, in turn,

$$5x^2 + 2x - 3 = 0,$$
$$(5x - 3)(x + 1) = 0,$$
$$x = 3/5 \quad \text{or} \quad x = -1.$$

We conclude that the solutions should be $3/5$ and -1. If we are suspicious of this we can always test these values in equation (i) to check that the equation is satisfied. For example, letting $x = 3/5$ gives $\sqrt{(x + 3)} = \sqrt{(18/5)} = 3\sqrt{(2/5)}$ while

$$\sqrt{(1 - x)} + \sqrt{(1 + x)} = \sqrt{(2/5)} + \sqrt{(8/5)} = 3\sqrt{(2/5)}$$

so that $x = 3/5$ is indeed a solution. A simpler calculation shows that $x = -1$ is also a solution, and we have completed the problem.

Now let us try equation (ii): $\sqrt{(x + 3)} = \sqrt{(1 - x)} - \sqrt{(1 + x)}$. Squaring gives $x + 3 = 1 - x - 2\sqrt{(1 - x)} \cdot \sqrt{(1 + x)} + 1 + x$ which we rearrange to give $x + 1 = -2\sqrt{(1 - x)} \cdot \sqrt{(1 + x)}$. Squaring again yields $x^2 + 2x + 1 = 4 - 4x^2$ which, as before, has solutions $x = 3/5$ and $x = -1$. If we now test $x = -1$ we obtain $\sqrt{(x + 3)} = \sqrt{2}$ while $\sqrt{(1 - x)} - \sqrt{(1 + x)} = \sqrt{2} - \sqrt{0} = \sqrt{2}$ so that $x = -1$ satisfies equation (ii). However, putting $x = 3/5$ gives $\sqrt{(x + 3)} = \sqrt{(18/5)} = 3\sqrt{(2/5)}$ while $\sqrt{(1 - x)} - \sqrt{(1 + x)} = \sqrt{(2/5)} - \sqrt{(8/5)} = \sqrt{(2/5)} - 2\sqrt{(2/5)} = -\sqrt{(2/5)}$ so in this case only $x = -1$ is a solution of (ii). Our method has produced one true solution and a spurious one.

If we consider equation (iii), $\sqrt{(x + 3)} = \sqrt{(1 + x)} - \sqrt{(1 - x)}$, and apply the same method we obtain, after squaring, the equation $x + 3 = 1 + x - 2\sqrt{(1 + x)} \cdot \sqrt{(1 - x)} + 1 - x$ which simplifies as in equation (ii) to give $x = 3/5$ or $x = -1$. In this case, testing $x = -1$ yields $\sqrt{(x + 3)} = \sqrt{2}$ and $\sqrt{(1 + x)} - \sqrt{(1 - x)} = -\sqrt{2}$ while putting $x = 3/5$ gives $\sqrt{(x + 3)} = 3\sqrt{(2/5)}$ and $\sqrt{(1 + x)} - \sqrt{(1 - x)} = \sqrt{(2/5)}$; both 'solutions' are spurious.

What is happening here? The method we have used to solve these equations is capable of introducing completely spurious numbers, so that we seem to need to check the answers. Since we do not normally need to check answers (except to correct the very human failing of making mistakes—and there are none above), why should we need to here? The resolution of this apparent difficulty is tackled in Chapter 2.

CHAPTER 2

Logic

In Chapter 1 we used some of the customary processes of mathematics to solve the three equations

(i) $\sqrt{(x+3)} = \sqrt{(1-x)} + \sqrt{(1+x)}$,
(ii) $\sqrt{(x+3)} = \sqrt{(1-x)} - \sqrt{(1+x)}$,
(iii) $\sqrt{(x+3)} = \sqrt{(1+x)} - \sqrt{(1-x)}$,

and we obtained a mixture of true solutions and spurious ones. To resolve this disturbing outcome we need to pay more attention to our methods.

In each case we start off with the equation, say equation (iii), $\sqrt{(x+3)} = \sqrt{(1+x)} - \sqrt{(1-x)}$. Then we square and carry on from there. Now, what we really mean at this step is that *if* $\sqrt{(x+3)} = \sqrt{(1+x)} - \sqrt{(1-x)}$ *then* $x+3 = 1 + x - 2\sqrt{(1+x)} \cdot \sqrt{(1-x)} + 1 - x$, and from the latter statement we deduce the next line and so on until we deduce $x = 3/5$ or $x = -1$. In other words we are noticing that *if* x satisfies the equation *then* $x = 3/5$ or $x = -1$. This cuts down the range of potential solutions from the whole set of real numbers to only two, and we can check individually whether or not these two potential solutions do actually satisfy the equation. The checking process establishes the truth or falsity of the statements '$x = 3/5$ is a solution' or '$x = -1$ is a solution', or, alternatively, such statements as 'if $x = 3/5$ then $\sqrt{(x+3)} = \sqrt{(1+x)} - \sqrt{(1-x)}$'.

It is obviously important to know when we need to use this checking process and when it is unnecessary, and for this we must sort out our logic. The main statements here are those of the form 'If A then B' where A and B are statements themselves. This is often stated as 'A implies B' and written $A \Rightarrow B$. Notice the meaning of this statement; it tells us that the two statements A and B are related in such a way that should A happen to be true then B must also be true. It does *not* indicate whether or not A or B themselves are true, nor does it give us any useful information about B in the case where A happens to be false.

From the two statements '$A \Rightarrow B$' and '$B \Rightarrow C$' we can deduce that '$A \Rightarrow C$'.
Going back to the examples above, we have, using implication signs:

$$\sqrt{(x+3)} = \sqrt{(1-x)} - \sqrt{(1+x)}$$
$$\Rightarrow x + 3 = 1 - x - 2\sqrt{(1-x)}\sqrt{(1+x)} + 1 + x$$
$$\Rightarrow x + 1 = -2\sqrt{(1-x)}\sqrt{(1+x)}$$
$$\Rightarrow x^2 + 2x + 1 = 4 - 4x^2$$
$$\Rightarrow 5x^2 + 2x - 3 = 0$$
$$\Rightarrow (5x-3)(x+1) = 0$$
$$\Rightarrow 5x - 3 = 0 \text{ or } x + 1 = 0,$$

where we mean here that each statement implies the one following it, so we
conclude that

$$\sqrt{(x+3)} = \sqrt{(1-x)} - \sqrt{(1+x)} \Rightarrow x = 3/5 \quad \text{or} \quad x = -1.$$

This is not exactly what we want, but it is a great step in the right direction,
since we have shown that at most two numbers are candidates for solutions
of the equation $\sqrt{(x+3)} = \sqrt{(1-x)} - \sqrt{(1+x)}$. We now need to check the
truth of the opposite implications, e.g. $x = -1 \Rightarrow \sqrt{(x+3)} = \sqrt{(1-x)} -$
$\sqrt{(1+x)}$. To check these is just arithmetic; notice that $x = 3/5$ does not imply
that $\sqrt{(x+3)} = \sqrt{(1-x)} - \sqrt{(1+x)}$.

There are, then, two possible implications connecting the statements A
and B: $A \Rightarrow B$ and $B \Rightarrow A$. It may happen that one of these is true without
the other (as in the paragraph above), or that both are true, or neither. Since
$B \Rightarrow A$ can be written just as neatly as $A \Leftarrow B$ we shall write $A \Leftrightarrow B$ to mean
'$A \Rightarrow B$ and $B \Rightarrow A$'. This may be restated as 'If A is true then B is true, and
if B is true then A is true'. $A \Leftrightarrow B$ therefore means that either A and B are
both true or they are both false, and is the most useful logical connection.
We usually pronounce $A \Leftrightarrow B$ as 'A (is true) if and only if B (is true)'.

It is important to distinguish between the three logical connections $\Rightarrow$, $\Leftarrow$
and $\Leftrightarrow$, the more so because everyday speech often blurs the distinction. In
the examples above, we squared numbers, and used, for example, the state-
ment $\sqrt{(x+3)} = \sqrt{(1-x)} - \sqrt{(1+x)} \Rightarrow x + 3 = 1 - x - 2\sqrt{(1-x)}\sqrt{(1+x)} +$
$1 + x$. This is as much as we can say immediately: if two numbers are equal
then their squares are equal. The converse is false since the squares of two
numbers may well be equal without the two numbers being equal; for example,
$2^2 = (-2)^2$.

Consider a simpler problem where all the connections are $\Leftrightarrow$:

$$x^2 + 2 = 2x + 1 \Leftrightarrow x^2 - 2x + 1 = 0$$
$$\Leftrightarrow (x-1)^2 = 0$$
$$\Leftrightarrow x - 1 = 0$$
$$\Leftrightarrow x = 1.$$

Therefore we have $x^2 + 2 = 2x + 1 \Leftrightarrow x = 1$, that is, if x satisfies the original
equation then $x = 1$ and if $x = 1$ then x satisfies the equation. In this case a

separate checking step is unnecessary as our original working has shown that x satisfies the equation if and only if $x = 1$. Therefore, to see whether a separate checking step is necessary we need to notice whether all the implications are $\Leftrightarrow$ or some are 'one way'.

All of the above assumes that we can decide whether or not one statement implies another. Since most mathematical statements involve implications, this is part of the mathematical repertoire; even a statement like $1 + 2 + \cdots + n = \frac{1}{2}n(n + 1)$ involves implications, since it is really a contracted form for 'If n is a positive integer, then the sum of the first n positive integers is $\frac{1}{2}n(n + 1)$'. Mathematics aims to prove that such general statements are true, thereby avoiding the need to test every case. We shall give various methods of proof in the next chapter and illustrate these by applying them to specific problems. Meanwhile let us start with some general ideas.

The most direct way of proving that $A \Rightarrow B$ (that is, proving the statement 'If A is true then B is true') is to assume the truth of A and deduce from that and any other information available (e.g. theorems already proved) that B must also be true.

Example Show that $\sqrt{(1 + x)} = 1 + \sqrt{(1 - x)} \Rightarrow x^2 = 3/4$. (It is tacitly assumed here that x is a real number. To be fully correct this ought to be said, but one can overdo correctness.)

Solution: Suppose that $\sqrt{(1 + x)} = 1 + \sqrt{(1 - x)}$.

Then

$$1 + x = 1 + 2\sqrt{(1 - x)} + 1 - x. \quad \text{(squaring)}$$
$$\therefore \quad 2x - 1 = 2\sqrt{(1 - x)}. \quad \text{(rearranging)}$$
$$\therefore 4x^2 - 4x + 1 = 4 - 4x. \quad \text{(squaring again)}$$
$$\therefore \quad 4x^2 = 3.$$
$$\therefore \quad x^2 = 3/4.$$

With the result written as above (using $\therefore$ to mean 'therefore') it is clear that the required conclusion $x^2 = 3/4$ is true, given our original supposition, so a concluding statement is not really necessary. In more complicated statements it may be necessary to sum up.

An alternative proof would be:

$$\sqrt{(1 + x)} = 1 + \sqrt{(1 - x)}$$
$$\Rightarrow \quad 1 + x = 1 + 2\sqrt{(1 - x)} + 1 - x \quad \text{(squaring)}$$
$$\Rightarrow \quad 2x - 1 = 2\sqrt{(1 - x)} \quad \text{(rearranging)}$$
$$\Rightarrow 4x^2 - 4x + 1 = 4 - 4x \quad \text{(squaring)}$$
$$\Rightarrow \quad 4x^2 = 3$$
$$\Rightarrow \quad x^2 = 3/4.$$

Again, the layout makes it clear that the first statement implies the last. The second form has the advantage that it is easy to check that the implications are all compatible; if we know that $A \Rightarrow B$ and $B \Leftarrow C$ then we cannot deduce any connection between A and C from these two. The second form is also more easily checked to see if we can immediately deduce the $\Leftrightarrow$ result (not in this case).

These two schemes of proving a result are more or less interchangeable. It is, however, worth noting that we may conclude that a statement is true purely as a consequence of the preceding line (as above), or it may be true as a consequence of several of the statements before it in combination. We shall reserve the $\Rightarrow$ sign for the former case (where one statement follows from the previous one), but allow 'therefore' to draw together various strands of an argument. This will become clear with practice.

Note: In the following example, and in the rest of mathematics, the word 'or' is taken to have its inclusive sense. Thus '$x = 0$ or $y = 0$' means that one or other or perhaps both of x and y are zero. If we ever wish to exclude the possibility of both being true, we shall say so explicitly.

Example   In each case below, decide on the most appropriate connection between A and B; x and y denote unknown real numbers and we wish the implications to be true for all real x and y:

1. $A:xy = 0$. $B:x = 0$ or $y = 0$.
2. $A:xy = 0$. $B:x = 0$ and $y = 0$.
3. $A:xy = 0$. $B:x \geq 0$ or $y \geq 0$.
4. $A:xy = 0$. $B:x^2 + y^2 \geq 1$.

In the first case, if $xy = 0$ then one or other (or maybe both) of x and y is zero, so $A \Rightarrow B$; also $B \Rightarrow A$ (if $x = 0$ or $y = 0$ then $xy = 0$) so here $A \Leftrightarrow B$. In the second example, if $x = 0$ and $y = 0$ then certainly $xy = 0$ so $A \Leftarrow B$; however, A does not imply B since it is quite possible for $xy = 0$ without both x and y being 0. For the third, $A \Rightarrow B$ (for if $xy = 0$ then either $x = 0$ or $y = 0$ whence either $x \geq 0$ or $y \geq 0$) but the opposite implication is false since B could be satisfied by $x = y = 1$ when A would be false. In the last example, neither statement implies the other.

When writing out proofs, the logical connections between the various statements must be displayed. This is necessary in the first place because *they are part of the proof* and also because this eliminates confusion between the various connections which could occur. Because the way in which we think up proofs is more intuitive (and less logical) than the finished result, there is a tendency to confuse one implication with another, and displaying the implications explicitly makes certain common mistakes much more glaring, and thus less likely to be made.

Caution: In establishing the truth of '$A \Rightarrow B$', we suppose A to be true and deduce things from that (since anything then deduced will be true if A is true). We *must not* assume B is true, that is, we must not assume the result we are trying to prove. To help avoid this blunder it is worth while ensuring that if the required statement is written down at the start as a reminder, then it is clearly marked as 'to be proved', or whatever, to prevent its subsequent misuse as if it had been proved.

Problems

1. Solve the following equations:
   (i) $\sqrt{(1+2x)} = 1 - \sqrt{x}$,   (ii)[†] $\sqrt{(1+x)} = 1 + \sqrt{(1-x)}$,
   (iii) $\sin x = \cos x$.

2. Insert the most appropriate logical conection ($\Rightarrow$, $\Leftarrow$, $\Leftrightarrow$ or none) between the statements of each pair:

   $$x = 3 \quad x^2 = 3x; \quad x^2 > 0 \quad x \neq 0;$$
   $$x^2 > 1 \quad x > 1; \quad x^2 + 2x + 1 = 1 \quad x = 0 \quad \text{or} \quad x = -2;$$
   $$x^2 + 2x + 1 = 1 \quad x = 0 \quad \text{or} \quad x = -1 \quad \text{or} \quad x = -2;$$
   $$x^2 + 2x + 1 = 1 \quad x = 0.$$

3. (i) By 'completing the square' (writing $x^2 + xy + y^2 = (x + \frac{1}{2}y)^2 +$ term not involving x) prove that $x^2 + xy + y^2 = 0 \Leftrightarrow x = 0$ and $y = 0$.
   (ii) Insert the most appropriate logical connection

   $$x^2 = y^2 \ldots x = y; \quad x^3 = y^3 \ldots x = y.$$

   [Prove those implications that are correct and give sample values of x and y to illustrate the falsity of those you consider wrong.]

---

† It may help to notice that $\{(1 + \sqrt{3})/2\}^2 = 1 + (\sqrt{3})/2$ and $\{(\sqrt{3} - 1)/2\}^2 = 1 - \sqrt{3}/2$.

CHAPTER 3

Proof

"That which is provable ought not to be believed in science without proof."

R. Dedekind

3.1 BEGINNINGS

We aim to produce results with all the certainty there is in mathematics. By an experimental approach we might test that the equation $1 + 2 + \cdots + n = n(n + 1)/2$ is true for a number of values of n and we might conjecture that it is true for all positive integers n. In mathematics we would hope to *prove* the truth of such a statement, even for those values of n which have not yet been individually tested.

To set our ideas straight, we shall start with some very ordinary results. Do not disdain these; their value is not in the results but in the method by which they are obtained. Simple results do not distract from the main technique as much as complicated ones.

To prove anything we need a starting point, that is, an agreed body of facts from which we deduce more complicated results; we shall assume the truth of the ordinary properties of arithmetic. These properties are so numerous that it would be extremely tedious to state them all; we may well use the result that $2 + 2 = 4$, for example. The properties we shall assume are those of addition, subtraction, multiplication, and division. To be precise, these are the results which follow from the assumption that there are two operations, called addition and multiplication, with the properties that the sum and product of two real numbers are real numbers and:

A1. For all real x, y and z, $x + (y + z) = (x + y) + z$ and $x(yz) = (xy)z$.
A2. For all real x and y, $x + y = y + x$ and $xy = yx$.
A3. There is a real number 0 such that, for all real x, $x + 0 = 0 + x = x$.
A4. For all real numbers x, there is a real number y such that
$x + y = y + x = 0$.
A5. There is a real number 1 such that, for all real x, $x1 = 1x = x$.
A6. For all real $x \neq 0$, there is a real y for which $xy = yx = 1$.
A7. For all real x, y and z, $x(y + z) = (xy) + (xz)$.
A8. $1 \neq 0$.

From these, we can, by the standard techniques of algebra, prove that if x is real, there is exactly one real number, denoted by $-x$, such that

$x + (-x) = 0$ and that $-(-x) = x$, and so on. Notice that nothing in properties A1–A8 can give us information about order properties, nor are such things as square roots mentioned.

In addition to the properties just stated, the real number system has an order, that is, some numbers are greater than others. To state this precisely: there is a relation which we shall denote by $<$ (and pronounce '$x < y$' as 'x is less than y'). For convenience we shall define $y > x$ to mean $x < y$ and define $x \leq y$ to mean $x < y$ or $x = y$. The assumptions we shall make about the order are:

For all real numbers x, y and z:

A9. Exactly one of the three statements $x < y$, $x = y$, $x > y$ is true.
A10. If $x < y$ and $y < z$ then $x < z$.
A11. If $x < y$ then $x + z < y + z$.
A12. If $x < y$ and $z > 0$ then $xz < yz$.

The order properties of numbers are not as straightforward as the arithmetical properties and they need some care. This is largely because most statements involving order are 'one way' implications whose converse is false (e.g. A10) and because the condition $z > 0$ in A12 causes trouble. For these reasons, more casual statements like 'add the inequalities' should be avoided in preference for more precise statements, as they tend to lead to errors. (We add the numbers, not the inequalities; this has practical significance when we wish to subtract quantities appearing in inequalities.)

We shall first establish some fairly obvious results which are not among our assumptions. This will set the scene and allow us to develop some techniques.

Example 1 If $x < y$ and $z < t$ then $x + z < y + t$.

Solution: Suppose that $x < y$ and $z < t$.
        Then $x + z < y + z$ (by A11, adding z to x and y).
        Also $z + y < t + y$ (by A11, adding y to z and t),
        whence $y + z < y + t$ (by A2).
        Therefore $x + z < y + z$ and $y + z < y + t$ so, by A10,
                    $x + z < y + t$.

Therefore $x < y$ and $z < t$ together imply that $x + z < y + t$, as required.

The last statement may be omitted where it is clear that the required result *has* been proved, but the writer should always check in his own mind that the correct conclusion has been reached.

Notice that a special case of Example 1, putting $x = z = 0$, yields $(y > 0$ and $t > 0) \Rightarrow y + t > 0$; the sum of two positive numbers is positive.

Example 2 $x < y \Leftrightarrow -y < -x$.

Note: Since none of the given properties A9–A12 nor Example 1 involves the $\Leftrightarrow$ connection, we must be prepared to establish each implication separately.

Solution: $x < y \Rightarrow x + (-x - y) < y + (-x - y)$ (A11)
$\quad\quad\quad \Rightarrow \quad\quad\quad -y < -x$ (arithmetic: A1–A4 in fact).

This establishes half of the required result. We could imitate this to prove the other half or we could notice that there is some symmetry. Since we have proved that, for all real x and y, $x < y \Rightarrow -y < -x$, this result will apply to the real numbers $-y$ and $-x$, so

$$-y < -x \Rightarrow -(-x) < -(-y) \Rightarrow x < y,$$

the last statement using the arithmetical result $-(-x) = x$.

Example 3 For all real numbers x, $x^2 \geq 0$, and if $x \neq 0$ then $x^2 > 0$.

Solution: Let x be a real number.
By A9, one of $x > 0$, $x = 0$, $x < 0$ is true.
If $x > 0$, then by A12 (with $z = x$) $\quad x^2 = x \cdot x > x \cdot 0 = 0$.
If $x = 0$, then $x^2 = 0$.
If $x < 0$, then by Example 2, $-x > -0 = 0$, so $(-x)^2 > 0$ (above), whence $x^2 = (-x)^2 > 0$.
Thus, in all cases, $x^2 \geq 0$, and if $x \neq 0$ then $x^2 > 0$.

The only significant point in Example 3 is the use of A9. The information that x is a real number tells us that it belongs to one (and only one) of three classes and we can treat each class in turn.

Example 4 If $x > 0$ then $1/x > 0$.

Solution: Suppose $x > 0$. Then $x \neq 0$ so a real number $1/x$ exists.
Since $1/x \neq 0$ (since $x \cdot (1/x) = 1$), $(1/x)^2 > 0$.
Since $x > 0$ and $(1/x)^2 > 0$, $1/x = x \cdot (1/x)^2 > 0$. (A12).

Example 5 If $0 < x$ and $x < y$ then $1/y < 1/x$.

Solution: Suppose that $0 < x$ and $x < y$.
$\therefore y > 0$. (By A10 since $0 < x$ and $x < y$.)
$\therefore xy > 0$. (By A12 since $0 < x$ and $y > 0$.)
$\therefore 1/(xy) > 0$. (Example 4.)
$\therefore x \cdot (1/(xy)) < y \cdot (1/(xy))$. (A12 with $z = 1/(xy)$.)
$\therefore 1/y < 1/x$.

Caution: It is easy to forget the condition $0 < x$ in Example 5!

These examples have been written out rather fully to show clearly why the various steps are valid. If a new result has to be established, students should prove it so that they are as certain of the new result as they are of the ones above. Notice that, however plausible emotive terminology like 'subtract the inequalities' may be, if $x < y$ and $z < t$ it does *not* follow that

$x - z < y - t$. (Consider $x = 1$, $z = 0$, $y = t = 2$.) An attempt to prove this might run:

Let $x < y$ and $z < t$.

Then $-t < -z$ (Example 2), so $x + (-t) < y + (-z)$ (Example 1).

$\therefore x - t < y - z$.

This result is the correct one, but notice it is not the one that loose chat might lead you to.

Notation: We write $a < x < b$ to mean $a < x$ and $x < b$. A similar interpretation applies to $a \leq x < b$ etc.

Problem: Prove that if $0 \leq a < b$ and $0 \leq c < d$ then $ac < bd$.

The greatest care with inequalities is needed when multiplication is involved, since the sign of the numbers may be crucial. This may involve splitting into cases.

Example 6 Let a and b be positive. Then $-a < x < b \Rightarrow x^2 < \max(a^2, b^2)$. (By $\max(a^2, b^2)$ we mean the maximum of a^2 and b^2.)

Solution: $-a < x < b \Rightarrow -a < x < 0$ or $0 \leq x < b$ (since $x < 0$ or $x \geq 0$)

$$\Rightarrow 0 < -x < a \quad \text{or} \quad 0 \leq x < b \text{ (Example 2)}$$
$$\Rightarrow (-x)^2 < a^2 \quad \text{or} \quad x^2 < b^2 \text{ (Problem above)}$$
$$\Rightarrow x^2 < a^2 \quad \text{or} \quad x^2 < b^2$$
$$\Rightarrow x^2 < \max(a^2, b^2) \quad \text{or} \quad x^2 < \max(a^2, b^2)$$
$$(\text{since } a^2 \leq \max(a^2, b^2), b^2 \leq \max(a^2, b^2))$$
$$\Rightarrow x^2 < \max(a^2, b^2).$$

Problem: Prove that if a and b are positive, $-a \leq x \leq b \Rightarrow x^2 \leq \max(a^2, b^2)$. (This should take you no more than a line or two!)

Example 7 Prove that $1 \leq x \leq 2 \Rightarrow 3/4 \leq x^2 - 3x + 3 \leq 1$.

In this case the problem is to determine bounds on the values of the function f, where $f(x) = x^2 - 3x + 3$, as x ranges over the interval between 1 and 2. We must not jump to the conclusion that $f(x)$ lies between $f(1)$ and $f(2)$, since we have no idea yet whether f 'preserves order' (i.e. whether $x < y \Rightarrow f(x) < f(y)$). In this case, indeed, $f(1) = f(2) = 1$ so if $f(1) \leq f(x) \leq f(2)$ for all x satisfying $1 \leq x \leq 2$, this would show that f is constant (which should look unlikely).

Method 1: Suppose that $1 \leq x \leq 2$.

Then $1 \leq x^2 \leq 4$ (since $0 < 1 < x \Rightarrow 1^2 < x^2$ etc.).   (1)

Also, $3 \leq 3x \leq 6$ so $-6 \leq -3x \leq -3$.   (2)

$\therefore 1 + (-6) \leq x^2 + (-3x) \leq 4 + (-3)$, (by (1), (2) and Example 1).

$\therefore \qquad -2 \leq x^2 - 3x + 3 \leq 4$.

Unfortunately, this quite correct deduction gives a conclusion which does not imply the result we wish; we must be more subtle. To see how to proceed, we notice that those values of x for which x^2 is close to its upper value in (1) are precisely those for which $-3x$ is far from its upper value in (2), so $x^2 - 3x$ is never close to $4 + (-3)$. We must try to deal with both terms simultaneously, which we do by 'completing the square', noticing that $x^2 - 3x = (x - 3/2)^2 - 9/4$.

Method 2: Suppose that $1 \leq x \leq 2$.
Then $-1/2 \leq x - 3/2 \leq 1/2$, so $(x - 3/2)^2 \leq \max((1/2)^2, (1/2)^2) = 1/4$ (by the last Problem). Also, $(x - 3/2)^2 \geq 0$ since it is a square.

$$\therefore 0 \leq x^2 - 3x + 9/4 \leq 1/4.$$
$$\therefore 3/4 \leq x^2 - 3x + 3 \leq 1.$$

Similar issues arise when reciprocals appear in expressions with inequalities.

Not all results will yield to a direct approach, so we must equip ourselves with one or two general methods of proof.

3.2 PROOF BY INDUCTION

We frequently wish to prove that a result is true for all positive integers. For example, if we consider adding the first n positive integers, $1 + 2 + 3 + \cdots + n$, we notice that the results for $n = 1, 2, 3, 4$ are $1, 3, 6, 10$ respectively, which we might (never mind how, for the moment) notice are the corresponding values of $\frac{1}{2}n(n + 1)$. How could we prove that, for all positive integers n, $1 + 2 + 3 + \cdots + n = \frac{1}{2}n(n + 1)$? Clearly, we can test as many individual values of n as we like, but we will never then have finished the job—there will always remain values of n that have not been tested. The basis for our method consists of:

The Principle of Induction

Suppose that $P(n)$ is some statement about the positive integer n. Then if

(i) $P(1)$ is true, and
(ii) for *every* positive integer k, $P(k) \Rightarrow P(k + 1)$,

we conclude that: for every positive integer n, $P(n)$ is true.

There are two things to notice. The statement (ii) will be true if we can show that for each positive integer k, $P(k) \Rightarrow P(k + 1)$, that is, *if $P(k)$ is true then $P(k + 1)$ must be true*. In other words, we need to show that the statement P is true for $k + 1$ if we are given the information that it is true for k (and that this deduction is valid for all k). Notice that this part is different from the conclusion, which tells us that $P(n)$ is true with no 'if' condition.

Secondly, notice that (ii) states that for every k the *implication* $P(k) \Rightarrow P(k + 1)$ holds. This is emphatically *not* the same as the statement $(P(k)$ is true for all $k) \Rightarrow P(k + 1)$; this statement is trivially true since if P is known to hold for all positive integers, it will hold for the integer $k + 1$. (The statement can also be criticised for using k with two meanings.)

The Principle of Induction is really a statement about the positive integers. If we know (i) and (ii) hold, then we know $P(1)$ is true, and by (ii) (with $k = 1$) we know $P(1) \Rightarrow P(2)$ hence $P(2)$ is true. Then, since $P(2) \Rightarrow P(3)$ we see that $P(3)$ is true. We could, apparently, proceed in this way to any particular case, but to say that $P(n)$ is true for all positive integers n amounts to the belief that every positive integer belongs to the sequence $1, 1 + 1, 1 + 1 + 1, \ldots$.

Example   Let $S(n) = 1 + 2 + \cdots + n$. Then, for all positive integers n,

$$S(n) = \tfrac{1}{2} n(n + 1).$$

Solution: Let $P(n)$ be the statement '$S(n) = \tfrac{1}{2} n(n + 1)$'.

(i) $P(1)$ is true, since $S(1) = 1 = \tfrac{1}{2} \cdot (1)(1 + 1)$.
(ii) Let k be a positive integer.
    Suppose $P(k)$ is true.
    Then $S(k) = \tfrac{1}{2} k(k + 1)$.
    $\therefore S(k + 1) = S(k) + (k + 1) = \tfrac{1}{2} k(k + 1) + k + 1 = \tfrac{1}{2}(k + 1)(k + 2)$.
    $\therefore P(k + 1)$ is true.

Therefore $P(k) \Rightarrow P(k + 1)$. (Summing up the last four lines.) Since the above was deduced only in the knowledge that k was a positive integer, it must be true for all such k. Therefore, for all positive integers k, $P(k) \Rightarrow P(k + 1)$. (We have established (ii) of the Principle of Induction.)

    $\therefore$ By induction, $P(n)$ is true for all positive integers n.

Pitfalls to Avoid: Notice that there is a distinction between an *expression*, like $S(n)$ above, and a *statement*, like $P(n)$ above. When an induction proof is written out, make sure your statement $P(n)$ is a statement *about n*. The statement 'For all positive integers n, $n^2 \geq 1$' is not about n, since the n there is a 'dummy' variable and could be replaced throughout by another symbol. The words 'for all....' introduce a 'dummy' variable.

The supposition we make in (ii) is about a particular (though unspecified) k, and we use this to deduce $P(k + 1)$. Only then do we observe that the implication $P(k) \Rightarrow P(k + 1)$ holds for all k.

Problems: (a) Prove that for every positive integer n,

$$1^2 + 2^2 + 3^2 + \cdots + n^2 = n(n + 1)(2n + 1)/6.$$

    (b) Prove that, for every positive integer n, $6^n - 1$ is a multiple of 5.

Definitions   At this point, since we shall be using the words 'for all' very often, we shall introduce the symbol $\forall$, an upside-down A, to mean this. This symbol is always followed by a symbol (a 'dummy' variable) and then a statement about that symbol.

As a further abbreviation we use standard notations for various sets of numbers. **N** denotes the set of positive integers (or natural numbers) consisting of 1, 2, 3, 4,.... We also use the notation $\in$ to mean 'belongs to' so that $n \in$ **N** means that n belongs to the set of natural numbers. Therefore Problem (a) above asks us to prove that $\forall n \in$ **N** $1^2 + 2^2 + \cdots + n^2 = n(n+1)(2n+1)/6$.

Statements using the symbol $\forall$ must be constructed grammatically so that they have a meaning in English. The symbol that follows the $\forall$ must be the dummy variable, nothing else making sense. The symbol used for the dummy variable is immaterial, but it must not duplicate a symbol used in the same sentence to mean something else. Thus '$\forall n \in$ **N** $n > k$' and '$\forall x \in$ **N** $x > k$' have the same meaning; we could not use k for the dummy variable, though.

The 'standard' form of induction we have used so far will prove that '$\forall n \in$ **N** $P(n)$' is true provided we can prove that, in symbols, $P(1)$ is true and $\forall k \in$ **N** $(P(k) \Rightarrow P(k+1))$. The step $P(k) \Rightarrow P(k+1)$ should not present undue difficulty in examples where $P(k)$ involves, say, the sum of the first k terms of some sequence. In other cases, however, the expressions involving $k+1$ may bear little relation to those involving k, so this method would be awkward. For example, the factors of $k+1$ bear no simple relationship to those of k, so if $P(k)$ is a statement about factors of k, our simple form of induction will not be easy to use. To avoid this difficulty we prove:

Theorem 3.1   Suppose that $P(n)$ is some statement about n and that

(i) $P(1)$ is true, and
(ii) $\forall k \in$ **N** $\{(P(1)$ and $P(2)$ and ... and $P(k)) \Rightarrow P(k+1)\}$.

Then, for all natural numbers n, $P(n)$ is true.

Proof: Let $Q(n)$ be the statement '$P(1)$ and $P(2)$ and ... and $P(n)$'. We shall apply ordinary induction to Q.
By (i) above, $Q(1)$ is true.
Let $k \in$ **N**. Suppose $Q(k)$ is true.
         Then, by (ii) $P(k+1)$ is true. $(Q(k) \Rightarrow P(k+1)$.)
         $\therefore (Q(k)$ and $P(k+1))$ is true, i.e. $Q(k+1)$ is true.
$\therefore Q(k) \Rightarrow Q(k+1)$.
Since k was merely supposed to belong to **N**, the last line holds for all $k \in$ **N**, i.e. $\forall k \in$ **N**, $Q(k) \Rightarrow Q(k+1)$.
Therefore by ordinary induction we deduce that $\forall n \in$ **N** $Q(n)$ is true. Since $Q(n) \Rightarrow P(n)$ we deduce that $\forall n \in$ **N** $P(n)$ is true.   □

Example   Every natural number greater than 1 is expressible as a product of prime numbers; we allow a prime number to be considered as the 'product' of one prime number.

[A natural number is said to be a *prime number* if it is larger than 1 and cannot be expressed as the product of two smaller positive integers. For technical reasons we do not call 1 a prime number. The first few primes are 2, 3, 5, 7, 11, 13, 17, 19,....]

Solution: The first issue is how to construct our statement P so that the exceptional number 1 is catered for. Let $P(n)$ be the statement 'Either $n = 1$ or n is a product of primes'.

Clearly $P(1)$ is true.

Let $k \in \mathbf{N}$ and suppose that $P(1), P(2), \ldots$ and $P(k)$ are all true. $k + 1$ is either a prime number (and hence the 'product' of one prime) or it is not a prime. If $k + 1$ is not a prime, then $k + 1 = ab$, where a and b are two natural numbers smaller than $k + 1$ (by definition). Thus a and b both lie between 2 and k (neither can be 1 or else the other must be $k + 1$). Since $P(1), \ldots, P(k)$ all hold, $P(a)$ and $P(b)$ are both true so both a and b are expressible as products of primes. (Remember $a \neq 1, b \neq 1$.) Therefore ab is a product of primes, hence so is $k + 1$.

$\therefore P(k + 1)$ is true.

Therefore by induction (as in Theorem 3.1), $P(n)$ is true for all $n \in \mathbf{N}$, so every integer greater than 1 is a product of primes.

Example   The Fibonacci numbers $a_1, a_2, a_3, \ldots$ are defined by setting $a_1 = a_2 = 1$ and $\forall n \geq 2$ $a_{n+1} = a_n + a_{n-1}$. Prove that

$$\forall n \in \mathbf{N} \quad a_n = \frac{1}{\sqrt{5}} \left\{ \left(\frac{1 + \sqrt{5}}{2} \right)^n - \left(\frac{1 - \sqrt{5}}{2} \right)^n \right\}.$$

Solution: Let $P(n)$ be

$$a_n = \frac{1}{\sqrt{5}} \left\{ \left(\frac{1 + \sqrt{5}}{2} \right)^n - \left(\frac{1 - \sqrt{5}}{2} \right)^n \right\}.$$

$P(1)$ is easily checked to be true, as is $P(2)$. (So the implication $P(1) \Rightarrow P(2)$ is true; think about it!) Now the formula $a_{n+1} = a_n + a_{n-1}$ can be used to show that, for all $n \geq 2$, $(P(1)$ and $\ldots$ and $P(n)) \Rightarrow P(n+1)$. Since we have shown this implication separately for $n = 1$, Theorem 3.1 will finish the job. The reader should fill in the details.

3.3 PROOF BY CONTRADICTION

At first sight this is an extremely perverse way of proving anything, but it turns out to be very useful.

Example   Let $n \in \mathbf{N}$ and n^2 be divisible by 2. Prove that n is divisible by 2.

Solution: Suppose the conclusion is false, i.e. suppose n is not divisible by 2.

By what we know of arithmetic, if we carry out long division of n by 2 there will be a non-zero remainder, which must be 1.

$\therefore$ For some integer $m \geq 0$, $n = 2m + 1$.

$\therefore$ For some integer $m \geq 0$, $n^2 = 4m^2 + 4m + 1 = 2(2m^2 + 2m) + 1$.

$\therefore n^2$ is not divisible by 2.

But this conclusion contradicts the given information (n^2 is divisible by 2) so our (unsupported) supposition that n is not divisible by 2 must be wrong, hence n is divisible by 2.

Remarks: The essence of this method of proof is to assume the required result is false and show that this is inconsistent with the given information. Care must be taken not to fall into the trap of assuming the truth of the desired result, for then nothing is achieved.

Proof by contradiction is useful when the conclusion is a 'negative' statement (one where more specific information is given by its being false) or where the required processes are easier to argue in the opposite direction from that required. In the example above, it is easier to deduce conclusions about n^2 given information about n than to deduce results about n with data about n^2.

Example For all *positive* x and y, $x^2 < y^2 \Rightarrow x < y$.

Solution: Recall from the Problem in Section 3.1 that $0 < x < y \Rightarrow x^2 < y^2$.

Suppose that x and y are positive and $x^2 < y^2$.

Assume $x \not< y$. Then by the order properties (A9), $x \geq y$. Thus $0 < y \leq x$, whence $y^2 \leq x^2$ (by the 'recalled' result, adding the case $x = y$). But this contradicts $x^2 < y^2$ so our assumption is untenable.

Therefore $x < y$.

We are now in a position to prove a famous result about numbers established by the ancient Greeks, and which caused much controversy at the time. A number is said to be **rational** if it is the ratio of two integers, so x is rational if there are two integers p and q such that $x = p/q$. If x can be represented in this way, then it will have many such representations since $p/q = 2p/2q = -p/-q$ etc. From the last of these three expressions we see that we may choose our representation so that its denominator is positive.

Theorem 3.2 There is no rational number whose square is 2.

Proof: Suppose there were a rational number whose square were 2. Call it $\sqrt{2}$.

Then there are two integers p and q, with $q > 0$, for which $\sqrt{2} = p/q$. If p and q are both divisible by 2, then there are integers p' and q' such that $p = 2p'$, $q = 2q'$ and $\sqrt{2} = p'/q'$. If p' and q' are both divisible by 2 we can once more remove a factor of 2 to obtain $\sqrt{2} = p''/q''$. We proceed in this way, extracting a factor of 2, as long as possible. The process must stop after a finite number of steps since $q'' < q' < q$ so after at most $q - 1$ steps the denominator cannot be further reduced.

Therefore, $\sqrt{2} = a/b$ where a and b are integers, $b > 0$ and a and b are not both divisible by 2. Therefore

$$a^2 = 2b^2.$$

$\therefore a^2$ is divisible by 2.

$\therefore a$ is divisible by 2. (First Example in this section.)

$\therefore$ For some integer a', $a = 2a'$ and $2b^2 = 4(a')^2$.

$\therefore 2(a')^2 = b^2$.

$\therefore b^2$, and hence b, is divisible by 2.

This is a contradiction, since we have proved that both a and b are divisible by 2 which we know to be false.

We have obtained a contradiction, so our initial assumption that $\sqrt{2}$ was rational is false, so there is no rational number whose square is 2. $\square$

Example   There are infinitely many prime numbers.

Solution: Suppose the result is false, that is, suppose there are only finitely many prime numbers. Call these prime numbers $p_1, p_2, \ldots, p_n$.

Consider the number $m = p_1 p_2 \ldots p_n + 1$. By an Example in Section 3.2, m has a prime factor, since it is expressible as a product of prime numbers. But none of the prime numbers $p_1, \ldots, p_n$ is a factor of m, since m leaves a remainder of 1 on division by each of $p_1, \ldots, p_n$, so the prime factor of m is another prime number, which contradicts the knowledge that $p_1, \ldots, p_n$ were all the primes.

The Lemma below is offered as an example of the use of induction and contradiction in combination. Its conclusion (that each non-empty set of natural numbers has a smallest element) may seem trivial and, if so, we ask the reader to bear with us until Chapter 6 when we consider sets of numbers in more detail. In fact, a set containing infinitely many real numbers need not have a smallest element, even if all its members lie between fixed bounds: the set $\{x : x \text{ is real and } 0 < x < 1\}$ has no smallest element. In comparison with more general sets of real numbers, the property of subsets of **N** which we are about to prove is special.

Lemma 3.3   Every non-empty set of natural numbers has a smallest element.

Proof: Let A be a non-empty set of natural numbers and suppose that it has no smallest element.

Let $P(n)$ denote the statement '$n \notin A$'.

Then $P(1)$ is true, since if $1 \in A$, 1 would be the smallest element of A. Also, if $1, 2, \ldots, n \notin A$ then $n + 1$ cannot belong to A, since then $n + 1$ would be the smallest element. Thus $(P(1)$ and $P(2)$ and $\ldots$ and $P(n)) \Rightarrow P(n + 1)$. Therefore, by induction, $\forall n \in \mathbf{N}$ $n \notin A$, so A contains no natural numbers. This is a contradiction, since A is supposed to be non-empty. $\square$

The use of proof by contradiction requires that we be able to negate statements. The negation of a statement is that statement which is true when the original statement is false and false when the original is true. The negation of complicated statements may be carried out step by step.

If 'A' and 'B' are two statements we may form the more complex statement 'A or B'. This statement is false exactly when 'A' and 'B' are both false, so if we denote the negation of a statement by 'not', then 'not $(A$ or $B)$' is equivalent to '(not A) and (not B)'. For example, the negation of '$x > 1$ or $x < 0$' is '$x \ngtr 1$ and $x \nless 0$' (which, for real x, we could more usefully write as '$x \leq 1$ and $x \geq 0$'). Similarly, the negation of 'A and B' is '(not A) or (not B)', so that the negation of '$x > 0$ and $y > 0$' is '$x \leq 0$ or $y \leq 0$'.

Negating the statement '$A \Rightarrow B$' is not so easy until we notice that it can be expressed in terms of the 'or' connective. The statement '$x = 1 \Rightarrow y = 2$' gives us no information about y if $x \neq 1$, and we see that it tells us that '$x \neq 1$ or $y = 2$'. Conversely, '$x \neq 1$ or $y = 2$' informs us that if $x = 1$ then $y = 2$, that is, '$x = 1 \Rightarrow y = 2$'. More generally, '$A \Rightarrow B$' is equivalent to 'A is false or B is true', that is, '(not A) or B'. (Suppose we know '(not A) or B' is true. Then, if 'A' is true, so that 'not A' is false, B must be true; that is, we know that 'if A is true then so is B'. Conversely, if we know that 'B is true if A is true', then either 'B' is true or 'A' must be false, giving '(not A) or B'.)

Let us now negate 'If p is a prime number and p divides ab, then either p divides a or p divides b'. This may be rewritten as

'(p is a prime and p divides ab) $\Rightarrow$ (p divides a or p divides b)'

so its negation is, simplifying in turn:

'not [(p is a prime and p divides ab) $\Rightarrow$ (p divides a or p divides b)]'

'not [not (p is a prime and p divides ab) or (p divides a or p divides b)]'

'(p is a prime and p divides ab) and not (p divides a or p divides b)'

'p is a prime and p divides ab and (p does not divide a and p does not divide b)'

Counterexamples

It is occasionally useful to prove certain results are false. This may arise when we wonder whether a condition in a theorem may be omitted, or it may arise when we are trying to prove some result and a plausible method involves a certain intermediate result—and knowing the falsity of the intermediate step may save time. Counterexamples showing that certain statements are false are also valuable in that they give insight—and promote caution!

Most mathematical results are of the general form 'for all x, x has a certain property' where the 'certain property' will usually involve various conditions. This statement is false if there is at least one x which fails to have the property in question. For example, to show that 'All odd integers greater than 1 are prime' is false we need only observe that there is an odd integer (e.g. 9) which is greater than 1 but not prime.

3.4 MORE INEQUALITIES

We start with an inequality which is simple, even crude. Its virtue lies in that it connects the operation of taking powers with the simpler one of multiplication of two numbers, and it does this simply.

Lemma 3.4 (Bernoulli's Inequality)   If $x \geq 1$ and $n \in \mathbf{N}$ then $x^n > n(x - 1)$.

Proof: By the Binomial Theorem,

$$x^n = ((x - 1) + 1)^n = \sum_{j=0}^{n} \binom{n}{j}(x - 1)^j 1^{n-j}.$$

Since all of the binomial coefficients are positive, and $(x-1) \geq 0$, each term in the sum is non-negative, so $x^n \geq \sum_{j=0}^{1} \binom{n}{j}(x-1)^j = 1 + n(x-1) > n(x-1)$. □

Example Let X be a natural number. Find an integer N such that, for all $n \geq N$, $3^n > X$.

Solution: By Bernoulli's Inequality, $3^n > 2n$, so $2n \geq X \Rightarrow 3^n > X$. Therefore if we set $N = X/2$ if X is even and $N = (X+1)/2$ if X is odd, then $n \geq N \Rightarrow 2n \geq X \Rightarrow 3^n > X$.

Remarks: Notice that there is no unique solution to this Example; if N_1 is chosen to be any integer greater than the N found above, then $n \geq N_1 \Rightarrow 3^n > X$. Bernoulli's Inequality allows us to find a simple expression for one of the many possible values of N for which $n \geq N \Rightarrow 3^n > X$. The penalty for the simplicity is that the value of N found by this process is usually greater than it need be; if $X = 1000$ the process above obtains $N = 500$, whereas $n \geq 7 \Rightarrow 3^n > 1000$. Bernoulli's Inequality is mainly useful in cases where a parameter, such as the X above, is involved.

A virtue of a simple result is that it can usually be adapted. This is the case with Bernoulli's Inequality where there are two immediate lines of attack. If $x \geq 1$ and $n \in \mathbb{N}$ then $x^n > n(x-1)$ and, since the right-hand side is non-negative, we may deduce that $(x^n)^2 > n^2(x-1)^2$. Thus $(x \geq 1$ and $n \in \mathbb{N}) \Rightarrow (x^2)^n > n^2(x-1)^2$. The gain is, of course, the n^2 appearing on the right, though to compensate we need to modify our value of x. To obtain information about 3^n we could choose $x = \sqrt{3}$ or perhaps a simpler number less than $\sqrt{3}$; thus $3^n > (\frac{9}{4})^n = (\frac{3}{2})^{2n} > n^2(\frac{1}{2})^2$. The other direction to adapt is in the case where $0 < x < 1$; then $1/x > 1$.

Problem: Show that if $0 < x < 1$ and $n \in \mathbb{N}$, $x^n < (1/n)(x/(1-x))$.

Bernoulli's Inequality will appear from time to time, but for the present we shall return to more central issues.

Definition If x is a real number, we define $|x|$, the **modulus** (or absolute value) of x, by

$$|x| = \begin{cases} x & \text{if } x \geq 0, \\ -x & \text{if } x < 0. \end{cases}$$

If x and y are two real numbers, we shall let $\max(x, y)$ denote the maximum of x and y, and $\min(x, y)$ denote their minimum.

From these two definitions we can see immediately that, for all real x, $|x| \geq 0$ and $|x| = \max(x, -x)$. Less obviously, notice that

$$\max(x, y) = [(x+y) + |x-y|]/2,$$

and

$$\min(x, y) = [(x+y) - |x-y|]/2.$$

Inequalities involving the modulus are simplified by the knowledge that $|x| \geq 0$ and will occur frequently. One key result is

Lemma 3.5 For all real numbers x and y, $|xy| = |x| \cdot |y|$ and $|x + y| \leq |x| + |y|$.

Proof: We prove these by looking at the various cases.

Case 1. If $x \geq 0$ and $y \geq 0$ then $xy \geq 0$ and $x + y \geq 0$ so

$$|xy| = xy = |x| \cdot |y| \quad \text{and} \quad |x + y| = x + y = |x| + |y|.$$

Case 2. If $x \geq 0$ and $y < 0$,

$$|xy| = -(xy) = x \cdot (-y) = |x| \cdot |y| \quad \text{(since } xy \leq 0\text{)}.$$

$$|x + y| = \begin{cases} x + y = |x| - |y| \leq |x| \leq |x| + |y| & \text{if} \quad x + y \geq 0, \\ -(x + y) = |y| - |x| \leq |y| \leq |x| + |y| & \text{if} \quad x + y \leq 0. \end{cases}$$

The remaining cases are similar. □

Remarks: The inequality $|x + y| \leq |x| + |y|$ is called the **triangle inequality**, and it is worth noticing that its right-hand side has a $+$ sign, always. If we wish to consider $|x - y|$ we notice that $|x - y| = |x + (-y)| \leq |x| + |-y| = |x| + |y|$, again with a $+$ sign. The temptation to put a $-$ sign on the right must be resisted as it may give false results. (Consider $x = 2$, $y = -2$.) The name 'triangle inequality' arises from the generalisation of this result to vectors where it can be geometrically interpreted.

Example $|x| < a \Rightarrow |xy| \leq a|y|$. (Notice that equality may occur on the right.)

Solution: If $|y| > 0$, then $|x| < a \Rightarrow |x| \cdot |y| < a|y|$ by the order properties. The only remaining possibility is $|y| = 0$ in which case $|x| \cdot |y| = a|y|$ ($= 0$).

Problem: Prove that $-1 < x < 1 \Leftrightarrow |x| < 1$.

We now come to an important type of problem. The underlying idea, which will emerge in later work, is that we wish to find out by how much we can allow x to vary from some given value without the values of some function of x varying by more than a prescribed amount. In the first case, we notice that $x^2 + 3x$ has the value 0 at $x = 0$ and we wish to find by how much we may let x vary from 0 and yet keep $x^2 + 3x$ within 1 of its value at 0.

Example Find a positive δ such that $|x| < \delta \Rightarrow |x^2 + 3x| < 1$.

Solution: Before starting, notice that we are only asked to find *a* positive value of δ with the required property, not the greatest possible one. We may therefore impose any convenient additional assumptions.

Suppose that $|x| < \delta$. Then

$$|x^2 + 3x| = |x| \cdot |x + 3| \leq \delta |x + 3| \quad \text{(above example since } |x| < \delta\text{)}$$
$$\leq \delta(|x| + |3|) \text{ (triangle inequality, noting } \delta > 0\text{)}$$
$$< \delta(\delta + 3)$$
$$\leq \delta(1 + 3) = 4\delta \quad \boxed{\text{IF } \delta \leq 1}$$
$$\leq 1 \qquad\qquad \boxed{\text{IF } \delta \leq 1/4}$$

Therefore, had we chosen δ in the first place to be positive and satisfy the two conditions in the boxes, we would obtain $|x| < \delta \Rightarrow |x^2 + 3x| < 1$, since one of the chain of inequalities is strict. Choosing a value of δ which satisfies the conditions in the boxes, we obtain

$$|x| < 1/4 \Rightarrow |x^2 + 3x| < 1.$$

The arbitrary assumption that $\delta \leq 1$ is done purely for convenience, since it avoids a more complicated calculation in finding a δ such that $\delta^2 + 3\delta \leq 1$. Had we supposed $\delta \leq 2$ here, instead, we would have obtained a different, but equally correct, set of constraints on δ. The key step, though, is that in which we obtain $|x^2 + 3x| \leq \delta|x + 3|$. From this point onwards it is enough to find a constant k, independent of x and δ, such that $|x| < \delta \Rightarrow |x + 3| < k$, whence $|x^2 + 3x| < k\delta$ and if we impose the additional constraint of $\delta \leq 1/k$ we shall have solved the problem.

If we do the same problem again, differently, we might obtain:

$$|x| < \delta \Rightarrow |x^2 + 3x| \leq \delta|x + 3| \leq \delta(|x| + 3) < \delta(\delta + 3)$$
$$\Rightarrow |x^2 + 3x| < (7/2)\delta \quad \boxed{\text{IF } \delta \leq 1/2}$$
$$\leq 1 \qquad\qquad \boxed{\text{IF } \delta \leq 2/7}$$

Thus, setting $\delta = 2/7$ we obtain $|x| < 2/7 \Rightarrow |x^2 + 3x| < 1$.

Problem: The reader is cordially invited to show that the largest value of δ for which $|x| < \delta \Rightarrow |x^2 + 3x| < 1$ is $(\sqrt{13} - 3)/2$. The volume of calculation will be a demonstration of the value of our method.

We may make the problem more useful by substituting a more general number on the right and considering x close to values other than 0.

Example   Given $\varepsilon > 0$ find a $\delta > 0$ for which

$$|x - 1| < \delta \Rightarrow |x^3 - 3x^2 - 2x + 4| < \varepsilon.$$

Solution: Let $|x - 1| < \delta$. Then

$$|x^3 - 3x^2 - 2x + 4| = |x - 1| \cdot |x^2 - 2x - 4| \leq \delta|x^2 - 2x - 4|$$
$$\leq \delta(|x|^2 + |2x| + |-4|)$$
$$\text{(triangle ineq.)}$$

Also, since $|x - 1| < \delta$, $|x| = |(x - 1) + 1| \leq |x - 1| + 1 < \delta + 1$, so

$$|x^3 - 3x^2 - 2x + 4| < \delta \cdot ((\delta + 1)^2 + 2(\delta + 1) + 4) \leq \delta \cdot (2^2 + 2.2 + 4)$$
$$\boxed{\text{IF } \delta \leq 1}$$

$$= 12\delta \le \varepsilon$$

$$\boxed{\text{IF } \delta \le \varepsilon/12}$$

Therefore $|x - 1| < \min(1, \varepsilon/12) \Rightarrow |x^3 - 3x^2 - 2x + 4| < \varepsilon$.

In this case, of course, we have not enough information to decide which of 1 or $\varepsilon/12$ is the smaller. Notice, again, the crucial step of showing that the required expression is less than a multiple of δ, and then that this multiplying factor is less than or equal to some constant.

Example   Let $\varepsilon > 0$. Find a positive δ such that $|x - 1| < \delta$ and $|y - 1| < \delta$ together imply that $|xy - 1|\varepsilon$.

Solution: Let $|x - 1| < \delta$ and $|y - 1| < \delta$. Then

$$|xy - 1| = |x(y - 1) + (x - 1)| \le |x| \cdot |y - 1| + |x - 1| < \delta(|x| + 1)$$
$$\le \delta(|x - 1| + 2) < \delta(\delta + 2)$$
$$\le 3\delta \quad \boxed{\text{IF } \delta \le 1}$$
$$\le \varepsilon \quad \boxed{\text{IF } \delta \le \varepsilon/3}$$

Therefore $|x - 1| < \min(1, \varepsilon/3)$ and $|y - 1| < \min(1, \varepsilon/3) \Rightarrow |xy - 1| < \varepsilon$.

Again, notice in the last example that we express the desired quantity, $|xy - 1|$, in terms of those which are 'small' in the sense that we can force such quantities to be as small as we need by a suitable choice of δ. In all these examples, one of the chain of inequalities is strict. If this were not so we would have to impose it at the end, e.g. $3\delta < \varepsilon$ (if $\delta < \varepsilon/3$), in which case we would have to be careful to name a particular δ which satisfies all the conditions—noticing that $\delta = \varepsilon/3$ does not satisfy $3\delta < \varepsilon$. (At this level it is such trivia that lead eventually to serious mistakes.)

In our final example, we need to cope with a denominator.

Example   Let $\varepsilon > 0$. Find a positive δ such that

$$|x| < \delta \Rightarrow \left| \frac{1}{1 + x} - 1 \right| < \varepsilon.$$

Discussion: The expression to be shown less than ε is $|-x/(1 + x)|$, so, assuming $|x| < \delta$, we see that this is less than $\delta/|1 + x|$. The usual technique is to prove that the factor multiplying δ is less than some constant (subject to suitable conditions). To show $1/|1 + x| \le K$ it is enough to prove $|1 + x| \ge 1/K$ (assuming K positive). So we need to find a condition of the form $|x| < \delta' \Rightarrow |1 + x| \ge K' > 0$ (putting K' for $1/K$). This is best spotted from a diagram (Fig. 3.1), since $|x| < \delta' \Rightarrow -\delta' < x < \delta' \Rightarrow 1 - \delta' < x < 1 + \delta'$.

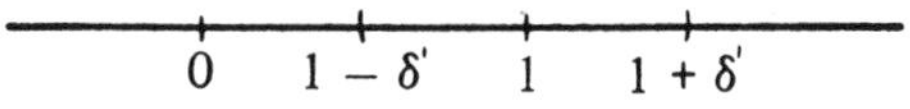

Fig. 3.1

For simplicity we shall choose δ' so that $1 - \delta' = 1/2$.

Solution: $|x| < \delta \Rightarrow -\delta < x < \delta \Rightarrow 1 - \delta < 1 + x \Rightarrow 1 + x > \frac{1}{2}$ $\boxed{\text{IF } \delta \leq \frac{1}{2}}$

$$\Rightarrow |1 + x| > \tfrac{1}{2}$$
$$\Rightarrow 1/|1 + x| < 2.$$

Let $|x| < \delta$. Then

$$\left| \frac{1}{1+x} - 1 \right| = \left| \frac{-x}{1+x} \right| \leq \frac{\delta}{|1+x|} < 2\delta \quad \boxed{\text{IF } \delta \leq \tfrac{1}{2}}$$

$$\leq \varepsilon \qquad \boxed{\text{IF } \delta \leq \varepsilon/2} \quad .$$

Therefore, $|x| < \min(\tfrac{1}{2}, \varepsilon/2) \Rightarrow |(1/(1+x)) - 1| < \varepsilon.$

Problems

1. Show that if $0 \leq x \leq a$ then $x^2 \leq a^2$.

2. Show that if $a > 0$ then $-a \leq x \leq a \Leftrightarrow x^2 \leq a^2$.

3. By observing that $x/(1+x) = 1 - 1/(1+x)$, or otherwise, show that $0 < x < y \Rightarrow 0 < x/(1+x) < y/(1+y)$.

4. Prove that $x < y \Leftrightarrow x^3 < y^3$. (Hint: Factorise $y^3 - x^3$.)

5. Given that $0 \leq x \leq 2$, prove that $2 \leq x^2 + 3x + 2 \leq 12$ and $-1/4 \leq x^2 - 3x + 2 \leq 2$.

6. Show that $x > 3 \Rightarrow x + 1/x > 10/3$. (Hint: Find the sign of $x + 1/x - 10/3$.)

7. Prove that, for all $n \in \mathbf{N}$, $1^3 + 2^3 + \cdots + n^3 = n^2(n+1)^2/4$.

8. Prove that, for all $n \in \mathbf{N}$, $3^{n+1} > (n+1)^2$ and hence, or otherwise, prove that for all $n \in \mathbf{N}$ $3^n > n^2$.

9. Define $a_n(n \in \mathbf{N})$ by $a_1 = 3/2$, $a_{n+1} = a_n^2 - 2a_n + 2$. Prove by induction that $\forall n \in \mathbf{N}$ $1 < a_n < 2$.

10. (i) Show that if x and y are positive, $\sqrt{x} - \sqrt{y} = (x - y)/(\sqrt{x} + \sqrt{y})$.
    (ii) Deduce that if $k \in \mathbf{N}$ then

    $$1/(2\sqrt{(k+1)}) \leq \sqrt{(k+1)} - \sqrt{k} \leq 1/\sqrt{(k+1)}.$$

    (iii) Define s_n by $s_n = 1 + 1/\sqrt{2} + \cdots + 1/\sqrt{n}$, and prove that $\forall n \in \mathbf{N}$, $\sqrt{n} \leq s_n \leq 2\sqrt{n}$.

11. Prove that if n is a natural number and n^2 is divisible by 3, then n is divisible by 3. Deduce that there is no rational number whose square is 3.

12. Prove that there is no rational number whose cube is 2.

13. Prove that if x is rational and y irrational then $x + y$ is irrational and, if $x \neq 0$, xy is irrational.

14. Prove that if $n \in \mathbf{N}$ and n^2 is divisible by 6 then n is divisible by 6 (you have already done the hard work!), and deduce that $\sqrt{6}$ is irrational. Show that if x^2 is irrational then x is irrational and hence, or otherwise, prove that $\sqrt{2} + \sqrt{3}$ is irrational.

15.* Show that if a and b are rational numbers such that $a\sqrt{2} + b\sqrt{3}$ is rational, then $a = b = 0$.

16.* Prove that if p and q are integers satisfying $p^3 + pq^2 + q^3 = 0$ then p is even if and only if q is even. Deduce that p and q are both even. Hence show that if x is a real number for which $x^3 + x + 1 = 0$ then x is irrational.

17.* In the usual decimal notation for an integer, $a_n a_{n-1} \ldots a_0$ denotes the number $10^n a_n + 10^{n-1} a_{n-1} + \cdots + 10 a_1 + a_0$, where $a_0, a_1, \ldots, a_n$ each represents an integer between 0 and 9. By showing that, for all $m \in \mathbf{N}$, $10^m - 1$ is a multiple of 9, prove that $10^n a_n + 10^{n-1} a_{n-1} + \cdots + a_0$ differs from $a_n + a_{n-1} + \cdots + a_0$ by a multiple of 9, and hence that a number is divisible by 9 if and only if the sum of its digits is divisible by 9.

18.* Prove that $10^n a_n + 10^{n-1} a_{n-1} + \cdots + a_0$ is divisible by 11 if and only if $a_0 - a_1 + a_2 - \cdots + (-1)^n a_n$ is divisible by 11.

19.* Define an operation $*$ on $\mathbf{N}$ by defining, for all $m, n \in \mathbf{N}$, $m*1 = m + 1$ and $m*(n + 1) = (m*n) + 1$. Prove the following:

   (i) $\forall m \in \mathbf{N}\ m*1 = 1*m$,
   (ii) $\forall m, n \in \mathbf{N}\ m*(n + 1) = (m + 1)*n$,
   (iii) $\forall m, n \in \mathbf{N}\ m*n = n*m$.

   [Hint: In (ii) let $P(n)$ be the statement '$\forall m \in \mathbf{N}\ m*(n + 1) = (m + 1)*n$'.] In a suitable context, this is the basis of a proof that addition of natural numbers is commutative, i.e. $x + y = y + x$.

20. Use Bernoulli's Inequality to prove that $(2/3)^n < 2/n$ if n is a natural number.

21. By noticing that $2 > (5/4)^3$, or otherwise, show that if $n \in \mathbf{N}$, $2^n > n^3/64$. Hence, given $\varepsilon > 0$, find a value N such that $n \geq N \Rightarrow n^2/2^n < \varepsilon$.

22. Prove that if x, y and z are real numbers, $|x + y + z| \leq |x| + |y| + |z|$.

23. Use induction to show that if $n \in \mathbf{N}$ and $a_1, \ldots, a_n$ are real numbers then $|a_1 + \cdots + a_n| \leq |a_1| + \cdots + |a_n|$.

24. Show that $(x \leq a$ and $-x \leq a) \Rightarrow |x| \leq a$.

25. By noticing that $x = (x - y) + y$, prove that if x and y are real numbers, $|x| - |y| \leq |x - y|$. Show also that $|y| - |x| \leq |x - y|$ and hence that $|\,|x| - |y|\,| \leq |x - y|$. This is a useful result worth remembering.

26. In each case below, find a positive δ for which the implication holds:

   (i) $|x| < \delta \Rightarrow |2x^2 + x| < 1$,
   (ii) $|x| < \delta \Rightarrow |x^3 + 5x^2 - 3x| < 1$,
   (iii) $|x| < \delta \Rightarrow |x^3 - 2x^2| < 1$.

27. Let $\varepsilon < 0$. Find, in each case, a positive δ with the required property:

   (i) $|x| < \delta \quad \Rightarrow |2x^2 + x| < \varepsilon$,

(ii) $|x| < \delta \quad \Rightarrow |x^3 + 5x^2 - 3x| < \varepsilon$,

(iii) $|x| < \delta \quad \Rightarrow |x^3 - 2x^2| < \varepsilon$,

(iv) $|x - 1| < \delta \Rightarrow |x^3 - 3x^2 + 2| < \varepsilon$,

(v) $|x + 2| < \delta \Rightarrow |x^3 - 2x + 4| < \varepsilon$.

28. Find a positive δ_1 such that $|x| < \delta_1 \Rightarrow |2 + x| \geq 1$. (You can either do this directly or use Q25, noting that $2 + x = 2 - (-x)$.) Hence find a positive δ_2 for which

$$|x| < \delta_2 \Rightarrow \left| \frac{x + 1}{x + 2} - \frac{1}{2} \right| < \varepsilon.$$

29. In each case below, find a positive δ such that the implication is true:

(i) $|x - 1| < \delta \Rightarrow \left| \frac{1}{x} - 1 \right| < \varepsilon$,

(ii) $|x - 1| < \delta \Rightarrow \left| \frac{x + 1}{x^2 + 1} - 1 \right| < \varepsilon$,

(iii) $|x + 1| < \delta \Rightarrow \left| \frac{x + 1}{x^2 + 2x} \right| < \varepsilon$.

30. Show that $|x + y| \geq 1 \Rightarrow |x| \geq \frac{1}{2}$ or $|y| \geq \frac{1}{2}$. More generally, prove that $|x_1 + x_2 + \cdots + x_n| \geq 1 \Rightarrow$ for some i with $1 \leq i \leq n$, $|x_i| \geq 1/n$.

31. There are many seemingly plausible results about moduli which are wrong. Give examples to show that, for arbitrary x, y and a,

$$x \leq y \not\Rightarrow |x| \leq |y|, \quad x \leq y \not\Rightarrow |x + a| \leq |y + a|,$$
$$|x + y| \leq 1 \not\Rightarrow |x| \leq 1 \quad \text{or} \quad |y| \leq 1.$$

CHAPTER 4

Limits

"...May we not call them the ghosts of departed quantities?"

G. Berkeley (1734)

The limit is one of the fundamental ideas of mathematics—and one of its most subtle. The history of the idea is long and fairly turbulent and it is only in the last century or so that a satisfactory theory has been established, mainly by Weierstrass and his pupils in the 1860s and 1870s.

Consider first the idea of a sequence, that is, an infinite progression of numbers $a_1, a_2, a_3, \ldots$. We shall denote such a sequence by (a_n), where by this we mean that for each natural number n there is a well determined nth element a_n of the sequence; n is a 'dummy' variable here and another symbol could be substituted. A sequence may be defined explicitly by a formula (e.g. $a_n = 2^n$) or by a recurrence relation such as $a_1 = 1$, $a_{n+1} = (\frac{1}{2})(a_n + 2/a_n)$ (for all $n \in \mathbf{N}$). For the second type of definition, we can determine a_{n+1} once we know a_n provided, in this case, $a_n \neq 0$ so we need to check, by induction, that a_n is defined for each $n \in \mathbf{N}$, that is, that the definition does not fail at some point. In the quoted example it is easy to see that $a_n > 0 \Rightarrow a_{n+1} > 0$ and that $a_1 > 0$, so, by induction, all terms are positive and no difficulty arises.

Consider the sequence $(1/n)$, whose nth term is $1/n$. Each term is positive but as n becomes larger, $1/n$ becomes smaller and 'approaches zero'. Indeed, $1/n$ can be made as small as we wish by choosing n sufficiently large and we think informally that 'in the limit' $1/n$ 'becomes zero'. It is this idea that we must make precise.

Definition Let (a_n) be a sequence of real numbers and let a be a real number. Then we say that $a_n \to a$ as $n \to \infty$ if and only if
For all $\varepsilon > 0$, there is a natural number N such that

$$\text{if} \quad n \geq N \quad \text{then} \quad |a_n - a| < \varepsilon.$$

In this case we say a is the **limit** of (a_n) and write $a = \lim_{n \to \infty} a_n$. The sign $\to$ is pronounced 'tends to'.

Remarks: 1. The definition demands that no matter what *predetermined* positive 'tolerance' ε we are given, there must be an integer N such that, for every $n \geq N$, the distance from a_n to a is less than ε. We may expect the suitable values of N to depend on ε.

2. ∞ is not a number. The word 'infinity' and the symbol ∞ are convenient terms to use to refer to this definition, but we have not given them a separate meaning.

3. Notice how we could show that $a_n \to a$ as $n \to \infty$. Choose ε to be positive and show that a suitable N exists, then observe that since we only use the fact that ε is positive, this must hold for all positive ε. The easiest way to show a value of N exists is to find one.

4. We use the symbol $\exists$ to mean 'there exists'. So $a_n \to a$ as $n \to \infty$ may be written, using s.t. to denote 'such that',

$$\forall \varepsilon > 0 \quad \exists N \in \mathbf{N} \quad \text{s.t.} \quad \forall n \geq N \quad |a_n - a| < \varepsilon. \qquad (*)$$

In this statement, ε, N and n are 'dummy' variables so that $(*)$ says something about the *sequence* (a_n) and the number a but not about ε, N or n. Notice that the order in which the symbols $\forall$ and $\exists$ appear is important; $(*)$ says that, no matter how $\varepsilon > 0$ is selected, once it has been selected, a suitable N exists. This is not at all the same as saying that a value of N exists which is suitable for all $\varepsilon > 0$.

5. There is nothing magic about the symbol ε. We may equally well rewrite $(*)$ as $\forall \delta > 0 \ \exists N \in \mathbf{N}$ s.t. $\forall n \geq N \ |a_n - a| < \delta$. Now choose $\varepsilon > 0$. Then, since $\varepsilon/2 > 0$, $(*)$ tells us that $\exists N_1$ s.t. $\forall n \geq N_1 \ |a_n - a| < \varepsilon/2$ and that $\exists N_2$ s.t. $\forall n \geq N_2 \ |a_n - a| < \varepsilon$. (Put $\delta = \varepsilon/2$, ε in turn.) We must be careful not to presume that the numbers N_1 and N_2, which we know exist, are equal, which is why we remind ourselves of this by using different suffices.

6. Notice that $|a_n - a| < \varepsilon \Leftrightarrow -\varepsilon < a_n - a < \varepsilon \Leftrightarrow a - \varepsilon < a_n < a + \varepsilon$. Therefore, $a_n \to a$ as $n \to \infty$ can also be written

$$\forall \varepsilon > 0 \quad \exists N \in \mathbf{N} \quad \text{s.t.} \quad \forall n \geq N \quad a - \varepsilon < a_n < a + \varepsilon.$$

This form is very often useful, especially when order properties are involved.

Example Let $a_n = 1/n$ for $n = 1, 2, 3, \ldots$. Then $a_n \to 0$ as $n \to \infty$. This, of course, should be no surprise, but we have made a technical definition of limit so we ought to check that it has the properties we might expect.

Solution: Let $\varepsilon > 0$. The problem is to show that there is a natural number N such that $\forall n \geq N \ |a_n - 0| < \varepsilon$. First, the rough work: $|a_n - 0| = 1/n$ and $1/n < \varepsilon$ if and only if $n > 1/\varepsilon$. Therefore all we have to do is choose $N > 1/\varepsilon$, then $n \geq N \Rightarrow n > 1/\varepsilon \Rightarrow |a_n - 0| < \varepsilon$. So, for the formal proof we have:
Let N be any integer greater than $1/\varepsilon$. Then:

$$\forall n \geq N \quad |a_n - 0| = 1/n \leq 1/N < \varepsilon.$$

Since ε was a typical positive number we have shown that this is true for all $\varepsilon > 0$, that is, $\forall \varepsilon > 0 \ \exists N \in \mathbf{N}$ such that $\forall n \geq N \ |a_n - 0| < \varepsilon$.
The proof just given raises an issue that it is easy to overlook: given a typical real number x, *is* there an integer $N > x$? (We used this tacitly above by choosing $N > 1/\varepsilon$—we need such an N to exist.) Now, our picture of the

real number system as lying along a line strongly suggests this, but we have
not yet proved it. For the time being, we shall set this aside by stating the
property we wish and making it one of our assumptions; we shall prove it
when we come to consider the structure of the real numbers in more detail
in Chapter 6.

Archimedean Property of the Real Numbers

If x is a real number, there is a natural number n for which $n > x$.

Example   For all $n \in \mathbf{N}$, let $a_n = a$. Then $\forall \varepsilon > 0 \; \exists N \in \mathbf{N}$ (e.g. $N = 1$) such that
$\forall n \geq N \; |a_n - a| = 0 < \varepsilon$. Thus $a_n \to a$ as $n \to \infty$.

Example   Let $0 < x < 1$ and, for all $n \in \mathbf{N}$, set $a_n = x^n$. Prove $a_n \to 0$ as $n \to \infty$.

Solution: We recall Bernoulli's Inequality from Section 3.4 in its second
form, giving in this case $x^n < (1/n) \cdot (x/(1 - x))$. Now

$$\frac{1}{n} \cdot \frac{x}{1 - x} \leq \varepsilon \Leftrightarrow n \geq \frac{1}{\varepsilon} \cdot \frac{x}{1 - x}$$

so we proceed as follows:
Let $\varepsilon > 0$. Choose N to be an integer with $N \geq x/((1 - x)\varepsilon)$.

$$\text{Then } \forall n \geq N, \; 0 < x^n < \frac{1}{n} \cdot \frac{x}{1 - x} \leq \frac{1}{N} \cdot \frac{x}{1 - x} \leq \varepsilon.$$

$$\therefore \forall n \geq N \quad |a_n - 0| = x^n < \varepsilon.$$

Since ε was arbitrary we have shown that

$$\forall \varepsilon > 0 \quad \exists N \in \mathbf{N} \quad \text{s.t.} \quad \forall n \geq N \quad |a_n - 0| < \varepsilon,$$

thus $a_n \to 0$ as $n \to \infty$.

Problem: Prove that if $-1 < x < 1$, then $x^n \to 0$ as $n \to \infty$.
   Armed with our formal definition, we need some rules for manipulating
limits. The most plausible results are:

Theorem 4.1   Suppose that (a_n) and (b_n) are sequences of real numbers, that
$a_n \to a$ and $b_n \to b$ as $n \to \infty$ and that λ is a real number. Then

   (i) $\lambda a_n \to \lambda a$ as $n \to \infty$;
   (ii) $a_n + b_n \to a + b$ as $n \to \infty$;
   (iii) $a_n b_n \to ab$ as $n \to \infty$.

(The statement $a_n + b_n \to a + b$ means that the sequence whose nth term is
$a_n + b_n$ tends to $a + b$ as $n \to \infty$.)

Proof: In all cases we have to show that for all $\varepsilon > 0$ there exists an N with certain properties. We shall do this by choosing an arbitrary positive ε and, using the information given, finding a suitable N. The procedure for thinking out these proofs is less formal than the proof itself, so we shall separate this from the proof. The left-hand column below would normally be omitted.

Rough Working

(i) Let $\varepsilon > 0$. We need to find $N \in \mathbf{N}$ so that

$$n \geq N \Rightarrow |\lambda a_n - \lambda a| = |\lambda|\,|a_n - a| < \varepsilon.$$

It would be enough to ensure that if $n \geq N$ then $|a_n - a| < \varepsilon/|\lambda|$, which is possible for $\lambda \neq 0$. If $\lambda = 0$, $|\lambda a_n - \lambda a| = 0 < \varepsilon$.

(ii) Let $\varepsilon > 0$. We must find N such that

$$\forall n \geq N \quad |(a_n + b_n) - (a + b)| < \varepsilon.$$

We can deduce something about $|a_n - a|$ and $|b_n - b|$, so we notice that

$$|(a_n + b_n) - (a + b)|$$
$$= |(a_n - a) + (b_n - b)|$$
$$\leq |a_n - a| + |b_n - b|$$

(by the triangle inequality). It would now be enough to find an N for which $\forall n \geq N$ both

$$|a_n - a| < \varepsilon/2 \quad \text{and} \quad |b_n - b| < \varepsilon/2.$$

Since $\varepsilon/2 > 0$, and $a_n \to a$ as $n \to \infty$

$$\exists N \quad \text{s.t.} \quad \forall n \geq N \quad |a_n - a| < \varepsilon/2.$$

Choose such a value of N and call it N_1, and, similarly, choose an N_2 such that

$$\forall n \geq N_2 \quad |b_n - b| < \varepsilon/2.$$

Formal Proof

(i) Suppose $\lambda \neq 0$. Let $\varepsilon > 0$. Then, since $\varepsilon/|\lambda| > 0$ and $a_n \to a$, there is an N s.t. $\forall n \geq N \; |a_n - a| < \varepsilon/|\lambda|$. For this N,

$$\forall n \geq N \quad |\lambda a_n - \lambda a| = |\lambda| \cdot |a_n - a|$$
$$< \varepsilon.$$

Since $\varepsilon > 0$ was arbitrary, this holds for all such ε, so

$$\forall \varepsilon > 0 \; \exists N \text{ s.t. } \forall n \geq N$$
$$|\lambda a_n - \lambda a| < \varepsilon.$$

$\therefore \lambda a_n \to \lambda a$ as $n \to \infty$.
If $\lambda = 0$, then $\forall n \in \mathbf{N}$
$\lambda a_n = 0 = \lambda a$ so the second Example of this chapter shows $\lambda a_n \to \lambda a$ as $n \to \infty$.

(ii) Let $\varepsilon > 0$. Since $\varepsilon/2 > 0$ and both $a_n \to a$ and $b_n \to b$ as $n \to \infty$, we can choose $N_1 \in \mathbf{N}$ and $N_2 \in \mathbf{N}$ such that

$$\forall n \geq N_1 \quad |a_n - a| < \varepsilon/2 \quad \text{and}$$
$$\forall n \geq N_2 \quad |b_n - b| < \varepsilon/2.$$

Now set $N = \max(N_1, N_2)$. Then $N \in \mathbf{N}$ and $n \geq N \Rightarrow n \geq N_1$ and $n \geq N_2$, so $\forall n \geq N$

$$|(a_n + b_n) - (a + b)|$$
$$\leq |a_n - a| + |b_n - b|$$
$$< \varepsilon/2 + \varepsilon/2 = \varepsilon.$$

Since $\varepsilon > 0$ was arbitrary, we have shown that

$$\forall \varepsilon > 0 \quad \exists N \in \mathbf{N} \quad \text{s.t.} \quad \forall n \geq N$$
$$|(a_n + b_n) - (a + b)| < \varepsilon.$$

Rough Working

(iii) Let $\varepsilon > 0$. We need to find an N s.t. $\forall n \geq N \ |a_n b_n - ab| < \varepsilon$. The information we have concerns $|a_n - a|$ and $|b_n - b|$, so we notice that

$$|a_n b_n - ab| = |a_n(b_n - b) + (a_n - a)b|$$
$$\leq |a_n| \cdot |b_n - b| +$$
$$|a_n - a| \cdot |b|. \qquad (*)$$

Since the RHS consists of more than one part we shall need care to ensure each part is small enough that their sum is less than ε.

At the moment we cannot see how small will be sufficient, so we introduce a positive constant k which we shall determine later once things become clearer. As in part (ii) we can find $N_0 \in \mathbf{N}$ such that

$$\forall n \geq N_0 \ |a_n - a| < k\varepsilon \text{ and}$$
$$|b_n - b| < k\varepsilon.$$

$\therefore \forall n \geq N_0$, using $(*)$,

$$|a_n b_n - ab| < |a_n| k\varepsilon + |b| k\varepsilon$$
$$= (|a_n| + |b|)k\varepsilon. \qquad (\dagger)$$

Again, there is a hitch. The RHS of $(\dagger)$ depends on n, so we cannot just let k be the reciprocal of $|a_n| + |b|$, since k is to be constant. Now,

$$|a_n| = |(a_n - a) + a| \leq |a_n - a| + |a|$$
$$< k\varepsilon + |a|.$$

We could use this, but comparing $(\dagger)$ with the inequalities in Chapter 3, it is simpler to obtain $|a_n| \leq C$ where C is a constant independent of ε. Now choose N_0' such that

$$\forall n \geq N_0' \ |a_n - a| < 1.$$

(Possible since $a_n \to a$ and $1 > 0$.)
Then if $n \geq N_0$ and $n \geq N_0'$,

$$|a_n b_n - ab| < (|a_n| + |b|)k\varepsilon$$
$$< (1 + |a| + |b|)k\varepsilon.$$

Formal Proof

(iii) Let $\varepsilon > 0$. Since $a_n \to a$ and $b_n \to b$ as $n \to \infty$, we can choose N_1, N_2 and N_3 all in $\mathbf{N}$ such that

$$\forall n \geq N_1 \ |a_n - a| < \varepsilon/(1 + |a| + |b|) \qquad (1)$$

$$\forall n \geq N_2 \ |b_n - b| < \varepsilon/(1 + |a| + |b|) \qquad (2)$$

$$\text{and } \forall n \geq N_3 \ |a_n - a| < 1. \qquad (3)$$

Let $N = \max(N_1, N_2, N_3)$, so $N \in \mathbf{N}$ and (since $n \geq N \Rightarrow n \geq N_1$ and $n \geq N_2$ and $n \geq N_3$)

$$\forall n \geq N$$
$$|a_n b_n - ab|$$
$$= |a_n(b_n - b) + (a_n - a)b|$$
$$\leq |a_n| \cdot |b_n - b| + |a_n - a| \, |b|$$
$$< (|a_n| + |b|)\varepsilon/(1 + |a| + |b|)$$
$$\leq (|a_n - a| + |a| + |b|)\varepsilon/(1 + |a| + |b|)$$
$$< (1 + |a| + |b|)\varepsilon/(1 + |a| + |b|)$$
$$= \varepsilon.$$

(The first and third inequalities by the triangle inequality, the second by (1) and (2) and the fourth by (3).)
Since $\varepsilon > 0$ was arbitrary we have shown that

$$\forall \varepsilon > 0 \quad \exists N \in \mathbf{N} \quad \text{s.t.} \quad \forall n \geq N$$
$$|a_n b_n - ab| < \varepsilon. \quad \square$$

Rough Working	Formal Proof
We now see that we should have chosen $k = 1/(1 + \|a\| + \|b\|)$.	

The ideas used in constructing the proofs above are more intuitive than the final proof—and for this reason it is essential to write out the final version neatly and clearly to check that the ideas in the outline do come together properly. (Not always, but that's life!) The most important points are the choice of the number $\varepsilon/(1 + |a| + |b|)$, which is made by working through the problem to see what is needed, and the choice of N_3 such that $\forall n \geq N_3\ |a_n - a| < 1$. As with the sort of inequality problem in Section 3.4, there are many suitable alternatives to the last; we could, for example, have chosen N_1 to satisfy $\forall n \geq N_1\ |a_n - a| < \min(1, \varepsilon)$.

In constructing proofs of limit results, notice that $a_n \to a$ as $n \to \infty$ if and only if $a_n - a \to 0$ as $n \to \infty$. This emphasises the nature of the information we are given by the statement $a_n \to a$ as $n \to \infty$; it is an elaborate statement about the 'smallness' of $a_n - a$. Thus to show $a_n b_n \to ab$ we must express $a_n b_n - ab$ in terms of 'small' quantities.

We know that $1/n \to 0$ as $n \to \infty$. By Theorem 4.1(i), $\alpha/n \to 0$ as $n \to \infty$ if α is a constant. By setting $a_n = b_n = 1/n$ for all $n \in \mathbf{N}$, (iii) shows us that $1/n^2 \to 0$ as $n \to \infty$. Using part (iii) again we could deduce $1/n^3 \to 0$ as $n \to \infty$ and so on. We shall leave it to the reader to prove that for all natural numbers k, $1/n^k \to 0$ as $n \to \infty$. (Induction?)

Theorem 4.2 Let (a_n) and (b_n) be two sequences of real numbers with $a_n \to a$ and $b_n \to b$ as $n \to \infty$. Then:

(i) $|a_n| \to |a|$ as $n \to \infty$;
(ii) If $a \neq 0$, $1/a_n \to 1/a$ as $n \to \infty$;
(iii) If $\forall n \in \mathbf{N}\ a_n \leq b_n$ then $a \leq b$.

Helpful Graffiti	Proof
(i) We use the inequality from Problem 3.25, $\big|\,|x| - |y|\,\big| \leq |x - y|$, so that if $|a_n - a| < \varepsilon$ then also $\big|\,|a_n| - |a|\,\big| < \varepsilon$.	(i) Let $\varepsilon > 0$. Then $\exists N \in \mathbf{N}$ s.t. $\forall n \geq N\ \ |a_n - a| < \varepsilon$. Using the alternative form of the triangle inequality we see that $$\forall n \geq N \quad \big|\,|a_n| - |a|\,\big| < |a_n - a| < \varepsilon.$$ Since $\varepsilon > 0$ was arbitrary, we have proved that $|a_n| \to |a|$ as $n \to \infty$.
(ii) Let $\varepsilon > 0$. Let k denote a positive constant whose value we shall fix once we see what we need. $\exists N_1$ such that $$\forall n \geq N_1 \quad |a_n - a| < k\varepsilon.$$ Then, $$\forall n \geq N_1 \quad \left|\frac{1}{a_n} - \frac{1}{a}\right| = \frac{|a - a_n|}{|a_n|\,|a|}$$	(ii) Let $\varepsilon > 0$. Since $a_n \to a$ as $n \to \infty$, we can choose N_1 and N_2 in $\mathbf{N}$ such that $$\forall n \geq N_1 \quad |a_n - a| < (|a|^2/2)\varepsilon \qquad (1)$$ and $\forall n \geq N_2 \quad |a_n - a| < |a|/2. \qquad (2)$ Let $N = \max(N_1, N_2)$. Then, by (2),

Helpful Graffiti

$$< \frac{k\varepsilon}{|a_n|\,|a|}.$$

The $|a|$ factor on the right could have been eliminated by a suitable choice of k, but the $|a_n|$ term is dependent on n. We want the RHS to be less than ε so it would be enough to ensure that $1/|a_n|$ is less than a constant for all sufficiently large n. This would be true if we ensured that $|a_n|$ was not close to zero. Since $a_n \to a$ as $n \to \infty$, $|a_n| \to |a| \neq 0$, so if we ensure a_n is 'close' to a, $|a_n|$ should be 'close' to $|a|$ and thus 'far' from 0. Choose N_2 s.t. $\forall n \geq N_2 \quad |a_n - a| < |a|/2$. Then, $\forall n \geq N_2 \quad |\,|a_n| - |a|\,| \leq |a_n - a|$
$$< |a|/2$$
so $|a| - |a|/2 < |a_n| < |a| + |a|/2$ thus $|a_n| > |a|/2$ and $1/|a_n| < 2/|a|$. Therefore, for $n \geq \max(N_1, N_2)$ we have

$$\left|\frac{1}{a_n} - \frac{1}{a}\right| = \frac{|a - a_n|}{|a|\,|a_n|} < \frac{k\varepsilon}{|a|} \cdot \frac{2}{|a|}$$

and so we see we should have chosen $k = |a|^2/2$.

(iii) We prove this by contradiction. Suppose $b < a$. Then in Fig. 4.1 for large enough n, b_n is close to b and a_n close to a. This should ensure $a_n > b_n$ which we know to be false. We need to choose $\varepsilon > 0$ small enough that no number can simultaneously be greater than $b + \varepsilon$ and less than $a - \varepsilon$.

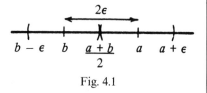

Fig. 4.1

Proof

$\forall n \geq N \quad |\,|a_n| - |a|\,| \leq |a_n - a|$
$$< |a|/2$$
so $|a|/2 = |a| - |a|/2 < |a_n|$.   (3)

$\therefore \forall n \geq N$

$$\left|\frac{1}{a_n} - \frac{1}{a}\right| = \frac{|a - a_n|}{|a_n|\,|a|}$$

$$< \frac{2}{|a|^2} \cdot |a_n - a| \quad \text{(by (3))}$$

$$< \frac{2}{|a|^2} \cdot \frac{|a|^2}{2}\varepsilon = \varepsilon \quad \text{(by (1))}.$$

Since ε was arbitrary,

$1/a_n \to 1/a \quad \text{as} \quad n \to \infty$.

Note: We have used the information $a \neq 0$ in three places: we divided by a, and in (1) and (2) the numbers on the RHS, $(|a|^2/2)\varepsilon$ and $|a|/2$, must be positive.

(iii) Suppose the result is false, that is, $a > b$. Then let $\varepsilon = (a - b)/2 > 0$ so, since $a_n \to a$ and $b_n \to b$ as $n \to \infty$, we can choose N_1 and N_2 in **N** such that

$\forall n \geq N_1 \quad a - \varepsilon < a_n < a + \varepsilon$ and
$\forall n \geq N_2 \quad b - \varepsilon < b_n < b + \varepsilon$.

Therefore letting $N = \max(N_1, N_2)$,

$b_N < b + \varepsilon = (a + b)/2 = a - \varepsilon < a_N$.

But we know that $\forall n \in \mathbf{N} \quad a_n \leq b_n$ so this is a contradiction, hence our supposition is wrong and

$a \leq b$. □

The proof in (iii) relies on our being able to find a natural number N for which $N \geq N_1$ and $N \geq N_2$. If, as in this case, N_1 and N_2 are natural numbers this is easy, but had they been real numbers we would have had to rely on the Archimedean property.

There is an issue here which we have avoided. If $a_n \to a$ as $n \to \infty$ and $a \neq 0$, then we showed in part (ii) above that there is a number N_2 with the property that $\forall n \geq N_2 \ |a_n| > |a|/2$ and hence, in particular, $\forall n \geq N_2 \ a_n \neq 0$. It may happen that there are values of n less than N_2 for which $a_n = 0$ and thus $1/a_n$ is meaningless. Strictly speaking, this means that the sequence $(1/a_n)$ is not properly defined. However, if we set $b_n = 1/a_n$ whenever $a_n \neq 0$ and $b_n = 0$ whenever $a_n = 0$, then by the proof above, $b_n \to 1/a$ as $n \to \infty$, since $\forall n \geq N_2$ $b_n = 1/a_n$. We shall presume this interpretation if the limit of $1/a_n$ is to be found.

Example   Find $\lim\limits_{n \to \infty} (n^2 + 2n)/(n^2 + 1)$.

Solution: We need to express this as a quotient of terms which tend to a limit, which we do by dividing numerator and denominator by the dominant term, n^2.

$$\frac{n^2 + 2n}{n^2 + 1} = \frac{1 + 2/n}{1 + 1/n^2}.$$

Since $1/n \to 0$, by Theorem 4.1(i) $2/n \to 0$ and 4.1(ii) tells us that $1 + 2/n \to 1$ as $n \to \infty$. Parts (ii) and (iii) show that $1 + 1/n^2 \to 1$, hence $1/(1 + 1/n^2) \to 1$ by 4.2(ii), so using 4.1(iii) again,

$$\frac{1 + 2/n}{1 + 1/n^2} \to \frac{1}{1} = 1 \quad \text{as} \quad n \to \infty$$

$\therefore \lim\limits_{n \to \infty} (n^2 + 2n)/(n^2 + 1) = \lim\limits_{n \to \infty} (1 + 2/n)/(1 + 1/n^2) = 1.$

Example   Find the limit as $n \to \infty$ of $\dfrac{n^3 + n + 1}{n^4 + 1}$.

Solution: Using the theorems several times,

$$\frac{n^3 + n + 1}{n^4 + 1} = \frac{1/n + 1/n^3 + 1/n^4}{1 + 1/n^4} \to \frac{0 + 0 + 0}{1 + 0} = 0 \quad \text{as} \quad n \to \infty.$$

If the denominator in an expression tends to 0 as $n \to \infty$, the theorems are no help, since the result about $\lim 1/a_n$ requires a_n to tend to a non-zero limit. Such cases are dealt with individually.

Example   Show that $(n^4 + 1)/(n^3 + 1)$ does not tend to a limit as $n \to \infty$.

Solution: $\dfrac{n^4 + 1}{n^3 + 1} = \dfrac{1 + 1/n^4}{1/n + 1/n^4}$,

so as $n \to \infty$ the numerator tends to 1, the denominator to 0. Now, for $n \in \mathbf{N}$,

$$\frac{n^4 + 1}{n^3 + 1} > \frac{n^4}{n^3 + 1} \geq \frac{n^4}{n^3 + n^3} = \frac{n}{2}$$

(since $n \geq 1$, so $n^3 + n^3 \geq n^3 + 1$). Now let $a_n = (n^4 + 1)/(n^3 + 1)$ and suppose $a_n \to a$ as $n \to \infty$. Then, since $1 > 0, \exists N \in \mathbf{N}$ such that $\forall n \geq N$ $a - 1 < a_n < a + 1$.
$\therefore \forall n \geq N$ $n/2 < a_n < a + 1$.
$\therefore \forall n \geq N$ $n < 2(a + 1)$, which is nonsense.
This contradiction shows that (a_n) does not tend to a limit.

Definition   A sequence (a_n) of real numbers is said to be **bounded above** if there is a constant U such that $\forall n \in \mathbf{N}$ $a_n \leq U$, and **bounded below** if there is a constant L such that $\forall n \in \mathbf{N}$ $a_n \geq L$. U and L are said to be **upper** and **lower bounds** respectively. (a_n) is said to be **bounded** if it is both bounded above and below.
   This allows us a general result including the last example as a special case.

Theorem 4.3   If (a_n) is a sequence of real numbers which tends to a limit, then (a_n) is bounded.

Proof: Suppose that $a_n \to a$ as $n \to \infty$. Since $1 > 0$, we can choose N such that $\forall n \geq N$ $a - 1 < a_n < a + 1$. Therefore if we let $M = \max(a + 1, a_1, \ldots, a_{N-1})$ and $m = \min(a - 1, a_1, \ldots, a_{N-1})$, we have $\forall n \in \mathbf{N}$ $m \leq a_n \leq M$. Thus (a_n) is bounded. $\square$

Remark: Notice the important issue here: the existence of the limit gives us information about all but finitely many values of n, so that in forming m and M we take the minimum or maximum of a *finite* number of quantities.
   The converse of Theorem 4.3 is false, as we shall see in a moment. Before tackling this, however, we must consider how to negate statements involving 'for all' or 'there exists'. Let $P(x)$ be some statement about x, and A be some set from which x is chosen; in our examples so far, A could be the set of natural numbers, or the set of natural numbers greater than N, etc. Then to say that 'For all x in A, $P(x)$ is true' is false, is equivalent to saying that 'There is an x in A for which $P(x)$ is false'; in symbols, the negation of '$\forall x \in A$ $P(x)$' is '$\exists x \in A$ s.t. not $P(x)$'. Similarly the negation of '$\exists x \in A$ s.t. $P(x)$' is '$\forall x \in A$ not $P(x)$'. For example, the negation of '$\forall x \in \mathbf{N}$ $x \geq 1$' is '$\exists x \in \mathbf{N}$ s.t. $x < 1$'.
   To negate a statement involving more than one $\forall$ or $\exists$ statement, we proceed in steps. The statement $a_n \nrightarrow a$ as $n \to \infty$ may be written in each of the following ways, the last being the most revealing:

$$\text{not} (\forall \varepsilon > 0 \quad \exists N \in \mathbf{N} \quad \text{s.t.} \quad \forall n \geq N \qquad |a_n - a| < \varepsilon)$$
$$\exists \varepsilon > 0 \quad \text{s.t.} \quad \text{not} (\exists N \in \mathbf{N} \quad \text{s.t.} \quad \forall n \geq N \quad |a_n - a| < \varepsilon)$$
$$\exists \varepsilon > 0 \quad \text{s.t.} \quad \forall N \in \mathbf{N} \text{ not } (\forall n \geq N \qquad |a_n - a| < \varepsilon)$$
$$\exists \varepsilon > 0 \quad \text{s.t.} \quad \forall N \in \mathbf{N} \exists n \geq N \quad \text{s.t.} \qquad |a_n - a| \geq \varepsilon.$$

Let us now use this to show that the sequence (a_n), where $a_n = (-1)^n$, does not tend to a limit. Let a be a real number; we show that $a_n \nrightarrow a$ as $n \to \infty$. If $a \geq 0$, set $\varepsilon = 1$ and we see that $\forall N \in \mathbf{N}$ $\exists n \geq N$ (e.g. $n = 2N + 1$) such that $|a_n - a| = |-1 - a| = a + 1 \geq \varepsilon$ so we have shown that $\exists \varepsilon > 0$ s.t. $\forall N \in \mathbf{N}$ $\exists n \geq N$ s.t. $|a_n - a| \geq \varepsilon$. If $a < 0$, let $\varepsilon = 1$ again and $\forall N \in \mathbf{N}$ $\exists n \geq N$ (e.g. $n = 2N$)

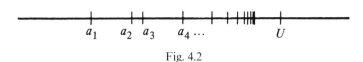

Fig. 4.2

s.t. $|a_n - a| = 1 - a \geq \varepsilon$. So in either case, a is not the limit of (a_n), so (a_n) has no limit. Notice, though, that (a_n) is bounded, since $\forall n \in \mathbf{N}$ $-1 \leq a_n \leq 1$, so the converse of Theorem 4.3 is false.

There is, however, something we wish to say in the way of a converse to Theorem 4.3. Consider a sequence (a_n) which is bounded above by U and in which each term is no smaller than the previous term, so $\forall n \in \mathbf{N}$ $a_{n+1} \geq a_n$ and $a_n \leq U$. Since the terms are increasing in size but never exceed U, Fig. 4.2 suggests that they must 'bunch up' in some sense and tend to a limit as $n \to \infty$. (Since U is just an upper bound and the sequence will have many upper bounds, there is no reason to expect the limit to be U.) This is quite a substantial statement since it can be thought of as saying that there are sufficiently many real numbers that there is one which is the limit of the sequence. In fact, this is a property that distinguishes the system of real numbers from the system of rational numbers, since an increasing sequence of rational numbers need not have a rational limit; the sequence 1, 1.4, 1.41, 1.414, ... of approximations to $\sqrt{2}$ has no rational limit.

We have not yet assumed enough about the real number system to prove the result we wish here, so we shall state it now and prove it once we have made a more detailed consideration of the real number system.

Definition   A sequence (a_n) of real numbers is said to be **increasing** if $\forall n \in \mathbf{N}$ $a_{n+1} \geq a_n$, and **strictly increasing** if $\forall n \in \mathbf{N}$ $a_{n+1} > a_n$. If $\forall n \in \mathbf{N}$ $a_{n+1} \leq a_n$ $(a_{n+1} < a_n)$ we say (a_n) is **decreasing (strictly decreasing)**.

Theorem 4.4   If (a_n) is an increasing sequence of real numbers which is bounded above, then (a_n) tends to a limit as $n \to \infty$.

The proof will appear in Chapter 6.

Corollary   If (a_n) is a decreasing sequence of real numbers which is bounded below, then (a_n) tends to a limit as $n \to \infty$.

Proof: Suppose that (a_n) is decreasing and $\forall n \in \mathbf{N}$ $a_n \geq L$. Let $b_n = -a_n$. Then $\forall n \in \mathbf{N}$ $b_{n+1} = -a_{n+1} \geq -a_n = b_n$, and $\forall n \in \mathbf{N}$ $b_n = -a_n \leq -L$ so (b_n) is increasing and bounded above. By the Theorem there is a b such that $b_n \to b$ as $n \to \infty$, so $a_n \to -b$ as $n \to \infty$. $\square$

Notice that the example $a_n = (-1)^n/n$ shows that a sequence may tend to a limit yet be neither increasing nor decreasing.

Example   Define a sequence (a_n) by $a_1 = 2$, $\forall n \in \mathbf{N}$ $a_{n+1} = (a_n + 2/a_n)/2$ and discuss whether or not it tends to a limit.

Solution: We first need to check that the definition is valid, that is, that a_n is not zero for any n or else the definition of a_{n+1} would fail. It is, however,

clear that $a_n > 0 \Rightarrow a_{n+1} > 0$ and, since $a_1 > 0$, induction shows that $\forall n \in \mathbf{N}$
$a_n > 0$.

To discover whether (a_n) is increasing or not, consider

$$a_{n+1} - a_n = \frac{1}{2}\left(a_n + \frac{2}{a_n}\right) - a_n = \frac{2 - a_n^2}{2a_n}. \tag{$*$}$$

To use $(*)$ we need to know the sign of $2 - a_n^2$. Now

$$2 - a_{n+1}^2 = 2 - \frac{1}{4}\left(a_n^2 + 4 + \frac{4}{a_n^2}\right) = -\frac{1}{4}\left(a_n^2 - 4 + \frac{4}{a_n^2}\right)$$

$$= -\frac{1}{4}\left(a_n - \frac{2}{a_n}\right)^2 \le 0,$$

so that $\forall n \in \mathbf{N}\ a_{n+1}^2 \ge 2$. Hence $\forall n \ge 2\ a_n^2 \ge 2$ and since we know $a_1^2 \ge 2$
we deduce that $\forall n \in \mathbf{N}\ a_n^2 \ge 2$. From $(*)$ we now see that $\forall n \in \mathbf{N}\ a_{n+1} - a_n \le 0$,
that is, (a_n) is decreasing. Since (a_n) is bounded below (all the terms are
positive) the Corollary above assures us that (a_n) tends to some real number,
say a, as $n \to \infty$.

We must find a. Since $a_n \to a$ as $n \to \infty$, $a_{n+1} \to a$ as $n \to \infty$ (see Problem 2).
Also $(a_n^2 > 2$ and $a_n > 0) \Rightarrow a_n > 1$, so $\forall n\ a_n \ge 1$, hence $a \ge 1$ (by Theorem 4.2(iii)
with $b_n = 1$ for all $n \in \mathbf{N}$). In particular $a \ne 0$, and

$$a = \lim_{n \to \infty} a_{n+1} = \lim_{n \to \infty} \frac{a_n^2 + 2}{2a_n} = \frac{a^2 + 2}{2a}.$$

From this we deduce that $a^2 = 2$, so $a_n \to a$, where a is a positive number
whose square is 2, i.e. $a = \sqrt{2}$.

Notice, in passing, that this result provides a proof that there is a real
number whose square is 2; the existence of a was deduced from Theorem
4.4. We shall return to this later.

When dealing with sequences defined by an equation of the form $a_{n+1} = f(a_n)$ it is often very helpful to draw a diagram. By plotting the two curves
$y = f(x)$ and $y = x$ on the same graph we can gauge the behaviour of (a_n) as
follows. Mark a_n on the x-axis, so that the point where the line $x = a_n$ meets
$y = f(x)$ has y-co-ordinate $f(a_n)$, that is, a_{n+1}. By drawing the line through
$(a_n, f(a_n))$ parallel to the x-axis until it meets the line $y = x$, we obtain a point
whose x-co-ordinate is a_{n+1}, and the process may be repeated; see Fig. 4.3.

When using diagrams in this way, care must be taken in the drawing and
in the correct placing of turning points and the like. Diagrams do, however,
often suggest clearly how a suitable analysis proof is to be made by suggesting
whether or not the sequence decreases, what suitable upper or lower bounds
are, etc. At this stage in your career you must then convert the ideas suggested
by the diagram into a formal proof, which will show up any significant
shortcomings in the sketch made.

In some examples we shall find that the terms of a sequence (a_n) become
large as n increases and are eventually greater than any given number. Since
(a_n) is not bounded in this case it cannot tend to a limit as $n \to \infty$, but we
can describe its behaviour.

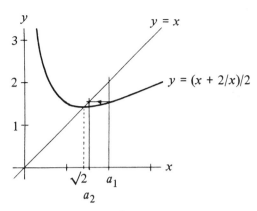

Fig. 4.3

Definition   We say that $a_n \to \infty$ as $n \to \infty$ if, for all real R, $\exists N \in \mathbf{N}$ such that $\forall n \geq N \; a_n > R$. We say that $a_n \to -\infty$ if $-a_n \to \infty$ as $n \to \infty$.

Notice that '$a_n \to \infty$ as $n \to \infty$' is a useful notation but we do not regard (a_n) as tending to a limit, nor do most of the results about limits apply. ∞ is not a real number and must not be treated as one.

Example   Show that $n(2 + (-1)^n) \to \infty$ as $n \to \infty$.

Solution: Let R be a real number. Choose N to be a natural number greater than R. Then $\forall n \geq N \; n(2 + (-1)^n) \geq n(2 - 1) = n \geq N > R$. Since R was arbitrary, this holds for all R and we have shown that $n(2 + (1)^n) \to \infty$.

The next two results, although 'elementary', require a certain amount of low cunning.

Example   Let $a > 0$. $a^{1/n} \to 1$ as $n \to \infty$.

Solution: We shall use Bernoulli's Inequality, since that will relate a to $a^{1/n}$, so suppose first that $a \geq 1$. Then $a^{1/n} \geq 1$ so, by Bernoulli's Inequality, $a > n(a^{1/n} - 1)$ whence $0 \leq a^{1/n} - 1 < a/n$, this being true for all $n \in \mathbf{N}$.

Let $\varepsilon > 0$. Choose $N \geq a/\varepsilon$. Then $\forall n \geq N \; |a^{1/n} - 1| < a/n \leq a/N \leq \varepsilon$. Since $\varepsilon > 0$ was arbitrary we have proved that $a^{1/n} \to 1$ as $n \to \infty$ (if $a \geq 1$).

For $0 < a < 1$, notice that $1/a > 1$ so $(1/a)^{1/n} \to 1$ as $n \to \infty$. Now $a^{1/n} = 1/(1/a)^{1/n} \to 1/1 = 1$ as $n \to \infty$.

Example $n^{1/n} \to 1$ as $n \to \infty$.

Solution: This one is more awkward, but illustrates how progress can be made. The same use of Bernoulli's Inequality as above would yield $n > n(n^{1/n} - 1)$ from which we deduce $n^{1/n} - 1 < 1$, which, unfortunately, does not imply that $n^{1/n} - 1$ is small for large n. We use the trick of using square roots, but in a different guise; apply Bernoulli's Inequality to $n^{1/2n}$ to obtain $\sqrt{n} = (n^{1/2n})^n > n(n^{1/2n} - 1)$, whence $0 \leq n^{1/2n} - 1 < 1/\sqrt{n}$.

Let $\varepsilon > 0$. Then $\forall n \geq N \geq 1/\varepsilon^2$,

$$|n^{1/2n} - 1| = n^{1/2n} - 1 < 1/\sqrt{n} \leq 1/\sqrt{N} \leq \varepsilon.$$

Thus $n^{1/2n} \to 1$ as $n \to \infty$, whence $(n^{1/n}) = (n^{1/2n})^2 \to 1$ as $n \to \infty$.

We have seen that the definition of limit we have introduced allows us to prove the results we would expect and, with the aid of the assumed properties of the real number system, also allows us to prove, without the aid of hand-waving, results like the existence of $\sqrt{2}$, which holds the promise of the technique's proving powerful. On the debit side there is no doubt that the definition and the basic results using it are technically complicated. This is necessary, as the next two examples illustrate, in the nature of warnings.

The definition of a limit demands that, as $n \to \infty$, a_n tends to a number, that is, something independent of n. It is tempting to write such things as '$a_n \to b_n$' to mean that a_n and b_n become arbitrarily close for large n, but the temptation should be resisted. Do we mean the *difference* becomes small, $a_n - b_n \to 0$? (True if $a_n = 1/n$, $b_n = 1/n^2$, for example.) Or do we mean that the *ratio* tends to 1, $a_n/b_n \to 1$? (True, for example, if $a_n = n^2 + 1$, $b_n = n^2 + n$.) Even worse, one is tempted to think that '$a_n \to b_n$' should imply '$a_n^2 \to b_n^2$' which is false in most interpretations; consider $a_n = n + 1/n$, $b_n = n$ (so that $a_n - b_n \to 0$ and $a_n/b_n \to 1$ as $n \to \infty$), but $a_n^2 = n^2 + 2 + 1/n^2$, $b_n^2 = n^2$ so $a_n^2 - b_n^2 \not\to 0$. Of course, if we notice $a_n - b_n \to 0$ we deduce, correctly, that $(a_n - b_n)^2 \to 0$, a different result. This is not profound, but it illustrates the sort of sloppiness that has to be avoided; indeed, in many problems the difficulty is in deciding precisely what the problem is, the solution being easy after that.

A second warning is more serious. Consider an argument along these lines: Let $f(m, n)$ depend on m and n and suppose $g(m) = \lim_{n \to \infty} f(m, n)$ exists, and that $g(m) \to G$ as $m \to \infty$. Thus G is close to $f(m, n)$ for sufficiently large m and n. If also $h(n) = \lim_{m \to \infty} f(m, n)$ and $h(n) \to H$ as $n \to \infty$, then $f(m, n)$ will be close to H for sufficiently large m and n. Does this mean $G = H$? Unfortunately not, as the following example shows.

$$\lim_{m \to \infty} \frac{m + n^2}{mn + n^2} = \lim_{m \to \infty} \frac{1 + n^2/m}{n + n^2/m} = \frac{1}{n} \quad \text{and} \quad \lim_{n \to \infty} \frac{1}{n} = 0.$$

Also

$$\lim_{n \to \infty} \frac{m + n^2}{mn + n^2} = \lim_{n \to \infty} \frac{m/n^2 + 1}{m/n + 1} = 1 \quad \text{and} \quad \lim_{m \to \infty} 1 = 1,$$

so

$$0 = \lim_{n \to \infty} \left\{ \lim_{m \to \infty} \frac{m + n^2}{mn + n^2} \right\} \neq \lim_{m \to \infty} \left\{ \lim_{n \to \infty} \frac{m + n^2}{mn + n^2} \right\} = 1.$$

The order in which limits are taken may not be reversed without justification. From time to time we shall find that we wish to reverse the order of taking certain limiting operations and we shall have to circumvent this difficulty.

Problems

1. Prove directly from the definition that:

   (i) if $a_n = (n-1)/(n+1)$ then $a_n \to 1$ as $n \to \infty$,
   (ii) if $\forall n \in \mathbf{N} \ 1 - (2/n^2) \le a_n \le 1 + 1/n$ then $a_n \to 1$ as $n \to \infty$,

   (iii) if $a_n = \begin{cases} 1/n & (n \text{ a perfect square}), \\ -1/n^2 & (\text{otherwise}), \end{cases}$ then $a_n \to 0$ as $n \to \infty$.

2. Suppose that (a_n) is a sequence of real numbers and that (b_n) is defined by $b_n = a_{n+1}$. Prove that $b_n \to L$ as $n \to \infty$ if and only if $a_n \to L$ as $n \to \infty$.

3. Let (a_n), (b_n) and (c_n) be three sequences of real numbers such that $\forall n \in \mathbf{N}$ $a_n \le b_n \le c_n$. Prove that if $a_n \to L$ and $c_n \to L$ as $n \to \infty$, then $b_n \to L$ as $n \to \infty$. (Sometimes called 'the sandwich rule'.)

4. Find the limit as $n \to \infty$, when one exists, or prove it does not exist, for the sequences whose nth term is:

   $$\frac{n^2 - n + 1}{2n^2 + n}, \quad \frac{n^4 + n^2 - 1}{3n^5 + 2}, \quad \frac{2^n + 3^n}{1 + 5^n}, \quad \frac{n^3 + n}{n^2 - 2}.$$

5. Prove that an increasing sequence is bounded if and only if it is bounded above.

6. Prove that (a_n) is bounded if and only if $(|a_n|)$ is bounded above.

7. Let (a_n) be an increasing sequence which is not bounded above. Prove that $a_n \to \infty$ as $n \to \infty$. Deduce that if (b_n) is increasing then either (b_n) tends to a limit or $b_n \to \infty$ as $n \to \infty$.

8. Define the sequence (a_n) by $a_1 = \alpha$ and $\forall n \in \mathbf{N}$ $a_{n+1} = a_n^2 - 2a_n + 2$, where $1 < \alpha < 2$. Show that (i) $\forall n \in \mathbf{N}$ $1 < a_n < 2$, (ii) (a_n) is decreasing, (iii) (a_n) tends to a limit, say a, as $n \to \infty$ and $a = 1$ or 2, (iv) $\forall n \in \mathbf{N}$ $a_n \le \alpha$ and $a \le \alpha$ and (v) $a_n \to 1$ as $n \to \infty$.

9. The sequence (a_n) is defined by $a_1 = \alpha$, $a_{n+1} = (a_n^2 - 2a_n + 3)/2$. Prove that if $1 \le \alpha < 3$ then $a_n \to 1$ as $n \to \infty$, while if $\alpha > 3$ then (a_n) is increasing and does not tend to a limit, hence $a_n \to \infty$ as $n \to \infty$ (see Q7). By considering a_2, discuss the behaviour of (a_n) in the cases in which $\alpha < 1$. (A diagram should help here.)

10. Define (a_n) by $a_1 = \alpha$, $a_{n+1} = (a_n^2 + a_n)/2$. Discuss the behaviour of (a_n) as $n \to \infty$ in the case where $0 \le \alpha < 1$. Discuss also the cases $\alpha = 1$, $\alpha > 1$, $-1 \le \alpha < 0$, $-2 < \alpha < -1$, $\alpha = -2$ and $\alpha < -2$. (A diagram may help see why these are the 'cases' to be considered.)

11. Let $a_1 = \alpha$, $a_{n+1} = 2a_n/(a_n + 1)$ and show that $a_n \to 1$ as $n \to \infty$ if $\alpha > 0$. How does the sequence behave if $\alpha < -1$? (Harder: Show that if $-1 < \alpha < 0$ and α is not of the form $-1/(2^m - 1)$ for some $m \in \mathbf{N}$, then $\forall n \in \mathbf{N}$ $a_n \ne -1$ so the sequence is well-defined and $a_n \to 1$ as $n \to \infty$.)

12. Let $a_1 = \alpha$, $a_{n+1} = 2a_n^2 - a_n^3$. Show that $a_n \to 0$ as $n \to \infty$ if $0 \le \alpha < 1$

and that $a_n \to 1$ as $n \to \infty$ if $1 \le \alpha \le (1 + \sqrt{5})/2$. (Hint: To show that $1 \le a_n \le (1 + \sqrt{5})/2 \Rightarrow 1 \le a_{n+1}$, consider $a_{n+1} - 1$ and factorise.)

13. Suppose that $a_1 = \alpha$ and $a_{n+1} = 2a_n/(1 + 2|a_n|)$. Find the behaviour of (a_n) as $n \to \infty$, depending on the value of α. (Sort out the cases yourself.)

14. Show that if $a_n \to \infty$ as $n \to \infty$ then $1/a_n \to 0$ as $n \to \infty$. Show also that if $\forall n \in \mathbf{N}$ $a_n > 0$ and $a_n \to 0$ as $n \to \infty$, then $1/a_n \to \infty$.

15. From Q14 it is easily seen that if $a_n \to \infty$ or $a_n \to -\infty$ as $n \to \infty$ then $1/a_n \to 0$ as $n \to \infty$. The converse is false; show this by finding a sequence (a_n) of non-zero numbers such that $1/a_n \to 0$ as $n \to \infty$ but a_n tends to neither ∞ nor $-\infty$. (Hint: Think of $|a_n|$.)

16. Let $a_n \to a$ as $n \to \infty$, and suppose that $\forall n \in \mathbf{N}$ $|b_n - a_n| < 1/n$. Prove that $b_n \to a$ as $n \to \infty$.

17.* Let $a_n \to a$ as $n \to \infty$ and define b_n by $b_{2n} = b_{2n-1} = a_n$ $\forall n \in \mathbf{N}$, so (b_n) is the sequence $(a_1, a_1, a_2, a_2, a_3, a_3, \ldots)$. Prove that $b_n \to a$ as $n \to \infty$. If (c_n) is the sequence $(a_1, a_2, a_2, a_3, a_3, a_3, \ldots)$ in which a_n occurs n times, show that $c_n \to a$ as $n \to \infty$.

18.* Let $a_n \to a$ and $b_n \to b$ as $n \to \infty$ and suppose that $\forall n \in \mathbf{N}$ $a_n < b_n$. Show by example that this information alone does not imply that $a < b$. Given that there is a constant $A > 0$ s.t. $\forall n \ge N$ $a_n + A \le b_n$, prove $a < b$.

19.* Theory tells us that if $a_n \to a$ then $a_n^2 \to a^2$ and $a_n^3 \to a^3$ as $n \to \infty$. In this question we investigate the converse. Any properties of square or cube roots used here must be proved true.

   (i) Show that if $a_n^2 \to 0$ as $n \to \infty$ then $a_n \to 0$ as $n \to \infty$.
   (ii) Show that if $a_n^2 \to a^2$ then $|a_n| \to |a|$ as $n \to \infty$. You may find the equation $x - y = (x^2 - y^2)/(x + y)$ helpful for the case $a \ne 0$.
   (iii) Give an example to show that it is possible that (a_n^2) tends to a limit while (a_n) does not.
   (iv) Prove that if $a_n^3 \to a^3$ as $n \to \infty$ then $a_n \to a$. (Hint: Find a suitable expression to use the idea in (ii).)
   (v) Prove that if $a_n^2 \to a^2$ and $a_{n+1} - a_n \to 0$ as $n \to \infty$ then either $a_n \to a$ or $a_n \to -a$ as $n \to \infty$. (Hint: For $a \ne 0$ show that $\exists N$ such that $\forall n \ge N$ a_n has constant sign.)

CHAPTER 5

Infinite Series

"And whatever the Ordinary Analysis performs using equations of a finite number of terms (when it be possible) this always performs using infinite equations; I have not hesitated to give it, too, the name of Analysis. Indeed, the arguments in it are no less certain than in the other, nor are the equations less exact, though we Mortals of finite intelligence can neither express nor conceive all their terms to know thence exactly the desired quantities."

Sir Isaac Newton

If we consider the sums $1 + 1/2, 1 + 1/2 + 1/4, 1 + 1/2 + 1/4 + 1/8, \ldots$ it is not hard to guess that $1 + 1/2 + \cdots + 1/2^k = 2 - 1/2^k$, which can then be proved for all $k \in \mathbf{N}$ by induction. Since $1/2^k \to 0$ as $k \to \infty$ we are inclined to think that if we added up all the numbers of the form $1/2^k$ for $k = 0, 1, 2, \ldots$ the sum should be 2. As there are infinitely many terms we cannot actually perform all the additions, but it is still clear what the answer 'should be'. This is what we have to make precise. Notice, however, that there are sequences whose terms cannot be added in this way; if we consider the sums $1, 1 + 1, 1 + 1 + 1, \ldots$ then it is clear that the sum of the first k terms does not tend to a limit as $k \to \infty$.

Definition   Let (a_n) be a sequence of real numbers and define s_n by $s_n = a_1 + a_2 + \cdots + a_n (n \in \mathbf{N})$. If (s_n) tends to a limit as $n \to \infty$ we say that **the series** $\sum a_n$ **converges**, while if (s_n) does not tend to a limit we say that $\sum a_n$ **diverges**. When $s_n \to s$ as $n \to \infty$ we say that s is the **sum** of the series and write $s = \sum_{n=1}^{\infty} a_n$.

The $\sum$ sign denotes summation, so that when we talk of 'the series $\sum a_n$' we are referring to the process of summing the terms a_n, that is, we are interested in the behaviour of the associated sequence (s_n). The statement '$\sum a_n$ converges' refers to the behaviour of (s_n) and does not directly say anything about (a_n). The notation $\sum_{n=1}^{\infty} a_n$ is used for a number, and its use implicitly requires the series $\sum a_n$ to converge since otherwise no such number exists. a_n is called the nth term of the series, and s_n is sometimes referred to as the nth partial sum.

Examples   (i) If $|x| < 1$ then $\sum x^n$ converges and $\sum_{n=1}^{\infty} x^n = x/(1-x)$.

Solution: For all $n \in \mathbb{N}$, $s_n = x + x^2 + \cdots + x^n = x(1-x^n)/(1-x)$ so, since $x^n \to 0$ as $n \to \infty$, $s_n \to x/(1-x)$ as $n \to \infty$.

(ii) $\sum 1/(n(n+1))$ converges and $\sum_{n=1}^{\infty} 1/(n(n+1)) = 1$.

Solution: Let $s_n = \sum_{k=1}^{n} 1/(k(k+1)) = \sum_{k=1}^{n} \{1/k - 1/(k+1)\} = 1 - 1/(n+1)$. $s_n \to 1$ as $n \to \infty$, by inspection.

The two examples just quoted are unusual in that we can find a convenient expression for s_n. Normally we have to prove that s_n tends to a limit by indirect means. The simplest result about series is:

Theorem 5.1   If $\sum a_n$ converges, then $a_n \to 0$ as $n \to \infty$.

Proof: Let $s_n = a_1 + a_2 + \cdots + a_n$, and suppose $\sum a_n$ converges. Then for some real number s, $s_n \to s$ as $n \to \infty$. Therefore $s_{n-1} \to s$ as $n \to \infty$ so that $a_n = s_n - s_{n-1} \to s - s = 0$ as $n \to \infty$. $\square$

Caution: The converse of Theorem 5.1 is false. There are sequences (a_n) for which $a_n \to 0$ as $n \to \infty$ but $\sum a_n$ diverges, as we shall show in a moment. This is a feature of the study of infinite series which leads to the diversity of the subject—there is no simple criterion which determines the convergence of a general series, just a host of partial results. Notice that Theorem 5.1 shows us that not all series converge.

Example   Let $a_n = 1/n \; \forall n \in \mathbb{N}$. Then $a_n \to 0$ as $n \to \infty$ but $\sum a_n$ diverges.

Solution: Let $s_n = 1 + 1/2 + 1/3 + \cdots + 1/n$ and choose m to be a non-negative integer for which $n \geq 2^m$. Then

$$s_n = 1 + \frac{1}{2} + \frac{1}{3} + \cdots + \frac{1}{2^m} + \cdots + \frac{1}{n}$$

$$\geq 1 + \frac{1}{2} + \left(\frac{1}{3} + \frac{1}{4}\right) + \cdots + \left(\frac{1}{2^{m-1}+1} + \cdots + \frac{1}{2^m}\right)$$

$$\geq 1 + \frac{1}{2} + \frac{1}{2} + \cdots + \frac{1}{2}$$

(since the bracket ending in $1/2^k$ has 2^{k-1} terms each no smaller than $1/2^k$, for $k = 1, \ldots, m$)

$$= 1 + m/2.$$

Therefore $s_{2^m} \geq 1 + m/2$ for all $m \in \mathbb{N}$, so if (s_n) were bounded above, by S say, then $\forall m \in \mathbb{N}$ $1 + m/2 \leq s_{2^m} \leq S$ which is a contradiction. It follows that (s_n) is not a bounded sequence and therefore does not tend to a limit as $n \to \infty$.

Before we develop the main body of theory there are some obvious little results:

Lemma 5.2 Let (a_n) and (b_n) be sequences of real numbers and λ be a constant. If $\sum a_n$ and $\sum b_n$ converge, so do $\sum \lambda a_n$ and $\sum (a_n + b_n)$ and $\sum_{n=1}^{\infty} \lambda a_n = \lambda \sum_{n=1}^{\infty} a_n$, $\sum_{n=1}^{\infty} (a_n + b_n) = \sum_{n=1}^{\infty} a_n + \sum_{n=1}^{\infty} b_n$.

Proof: Let $s_n = a_1 + a_2 + \cdots + a_n$ and $t_n = b_1 + b_2 + \cdots + b_n$. Then if $s_n \to s$ and $t_n \to t$ as $n \to \infty$

$$\lambda a_1 + \lambda a_2 + \cdots + \lambda a_n = \lambda s_n \to \lambda s \quad \text{as} \quad n \to \infty$$

and

$$(a_1 + b_1) + (a_2 + b_2) + \cdots + (a_n + b_n)$$
$$= s_n + t_n \to s + t \quad \text{as} \quad n \to \infty. \quad \square$$

Lemma 5.3 Suppose that there is a natural number N for which $\forall n \geq N \ a_n = b_n$. Then $\sum a_n$ and $\sum b_n$ either both converge or both diverge.

Proof: Let $s_n = a_1 + \cdots + a_n$, $t_n = b_1 + \cdots + b_n$. Then $\forall n \geq N \ s_n = t_n + C$ where $C = a_1 + a_2 + \cdots + a_{N-1} - b_1 - b_2 - \cdots - b_{N-1}$ is independent of n. (s_n) tends to a limit if and only if (t_n) tends to a limit (though the two limits will differ by C). $\quad \square$

Remark: Changing the first few terms of a series does not affect the convergence—but it will normally affect the sum.

We now start with series of non-negative terms, whose behaviour is simpler than the general case.

Lemma 5.4 Let $\sum a_n$ be a series with non-negative terms. Then $\sum a_n$ is convergent if and only if the associated sequence (s_n) is bounded, that is, if and only if there is a constant K for which $\forall N \in \mathbf{N} \ \sum_{n=1}^{N} a_n \leq K$.

Proof: $\forall n \in \mathbf{N} \ s_{n+1} = s_n + a_{n+1} \geq s_n$ (since $\forall k \ a_k \geq 0$), so (s_n) is increasing. Therefore from Theorem 4.4, (s_n) tends to a limit if and only if it is bounded above. $\quad \square$

This leads to our first standard test for convergence. Notice that, very roughly, the test says that a series of non-negative terms converges if $a_n \to 0$ 'sufficiently rapidly'.

Lemma 5.5 (The Comparison Test) Suppose that $\forall n \in \mathbf{N} \ 0 \leq a_n \leq b_n$ and that $\sum b_n$ converges. Then $\sum a_n$ converges.

Proof: Let $s_n = a_1 + a_2 + \cdots + a_n$ and $t_n = b_1 + b_2 + \cdots + b_n$. Then, since $\sum b_n$ converges, (t_n) is bounded. Let $t_n \leq T \ \forall n \in \mathbf{N}$. Then, $\forall n \in \mathbf{N}$,

$$s_n = a_1 + a_2 + \cdots + a_n \leq b_1 + b_2 + \cdots + b_n = t_n \leq T$$

so (s_n) is bounded above. Since (s_n) is increasing $(a_n \geq 0)$, (s_n) tends to a limit as $n \to \infty$, and $\sum a_n$ converges. $\quad \square$

Corollary 1 Suppose that $N \in \mathbf{N}$, $\forall n \geq N \ 0 \leq a_n \leq b_n$ and $\sum b_n$ converges. Then $\sum a_n$ converges.

Proof: Let $a'_n = b'_n = 0\,(n = 1, 2, \ldots, N - 1)$ and $a'_n = a_n \;\; \forall n \geq N, \;\; b'_n = b_n$ $\forall n \geq N$. By Lemma 5.3, $\sum b'_n$ converges, so by the comparison test $\sum a'_n$ converges and Lemma 5.3, once more, tells us that $\sum a_n$ converges. $\quad\square$

Corollary 2 Suppose that $N \in \mathbf{N}$, that $\forall n \geq N \;\; 0 \leq c_n \leq d_n$ and that $\sum c_n$ diverges. Then $\sum d_n$ diverges.

Proof: If $\sum d_n$ were convergent, Corollary 1 would show $\sum c_n$ convergent. $\quad\square$

Remarks: These results rely on the terms being non-negative, at least for large n. We can use them to show that $\sum 1/n^2$ converges as follows: $\sum 1/n(n + 1)$ we know converges, hence $\sum 2/n(n + 1)$ converges, from which, noticing that $\forall n \in \mathbf{N} \;\; 1/n^2 \leq 2/n(n + 1)$, the comparison test shows $\sum 1/n^2$ converges. By noticing that $\forall n \in \mathbf{N} \;\; 0 \leq 1/n \leq 1/\sqrt{n}$ we see that $\sum 1/\sqrt{n}$ diverges, since $\sum 1/n$ does.

Corollary 3 Suppose that $\forall n \in \mathbf{N} \;\; a_n \geq 0$ and $b_n > 0$ and that $a_n/b_n \to \lambda$ as $n \to \infty$. Then if $\sum b_n$ is convergent, so is $\sum a_n$.

Proof: Since $a_n/b_n \to \lambda$, $\exists N$ s.t. $\forall n \geq N \;\; \lambda - 1 < a_n/b_n < \lambda + 1$. Therefore, $\forall n \geq N \;\; a_n < (\lambda + 1)b_n$ and, if $\sum b_n$ is convergent, $\sum (\lambda + 1)b_n$ is also, whence Corollary 1 shows that $\sum a_n$ converges. $\quad\square$

Remarks: This Corollary is not quite symmetric in (a_n) and (b_n). If $a_n/b_n \to \lambda$ as $n \to \infty$, then provided $\lambda \neq 0$, $b_n/a_n \to 1/\lambda$ as $n \to \infty$ so the two series will either both converge or both diverge in the case $\lambda \neq 0$. If $\lambda = 0$ it may happen that $\sum a_n$ converges and $\sum b_n$ diverges.

The drawback of all these comparison tests is that we need to have a stock of series at our disposal to compare new series with. To establish some we need one rather specialised result.

Theorem 5.6 (Cauchy's Condensation Test) Suppose that (a_n) is a *decreasing* sequence of positive terms. Then $\sum a_n$ converges if and only if $\sum 2^n a_{2^n}$ converges.

Proof: Let $s_n = a_1 + a_2 + \cdots + a_n$ and $t_n = 2^1 a_2 + 2^2 a_4 + \cdots + 2^n a_{2^n}$. Since (a_n) is decreasing and $a_1 \geq 0$

$$s_{2^n} = a_1 + a_2 + (a_3 + a_4) + \cdots + (a_{2^{n-1}+1} + \cdots + a_{2^n})$$
$$\geq 0 + (a_2) + (a_4 + a_4) + \cdots + (a_{2^n} + \cdots + a_{2^n})$$

where the kth bracket contains 2^{k-1}

occurrences of a_{2^k}

$$\geq a_2 + 2a_4 + \cdots + 2^{k-1} a_{2^k} + \cdots + 2^{n-1} a_{2^n}$$
$$= \tfrac{1}{2} t_n. \tag{1}$$

Also, $s_{2^n} = a_1 + (a_2 + a_3) + (a_4 + \cdots + a_8) + \cdots$

$\qquad\qquad + (a_{2^{n-1}} + \cdots + a_{2^n-1}) + a_{2^n}$

$\qquad \le a_1 + (a_2 + a_2) + (a_4 + \cdots + a_4) + \cdots$

$\qquad\qquad + (a_{2^{n-1}} + \cdots + a_{2^{n-1}}) + a_{2^n}$

where the kth bracket contains 2^k occurrences of a_{2^k}

$\qquad = a_1 + 2a_2 + 4a_4 + \cdots + 2^{n-1}a_{2^{n-1}} + a_{2^n}$

$\qquad \le a_1 + t_n. \qquad\qquad\qquad\qquad\qquad\qquad\qquad\qquad (2)$

If the sequence (s_k) is bounded, then (s_{2^n}) is bounded (since its terms are a selection of those of (s_k)) and so by (1), (t_n) is bounded above. Also if (t_n) is bounded above, by T say, then (2) shows that

$$\forall n \in \mathbf{N} \quad s_n \le s_{2^n} \le a_1 + t_n \le a_1 + T,$$

(the first inequality because $n < 2^n$ and the extra terms in s_{2^n} are positive) whence (s_n) is bounded above. Thus (s_n) is bounded above if and only if (t_n) is bounded above, and since all terms in both series are positive, Lemma 5.4 completes the proof. $\quad\square$

Example $\sum 1/n^\alpha$ converges if $\alpha > 1$ and diverges if $\alpha \le 1$.

Solution: We need only consider $\alpha > 0$, since if $\alpha \le 0$ then $1/n^\alpha \nrightarrow 0$, so the series certainly diverges. Let $a_n = 1/n^\alpha$ and $\alpha > 0$. Then, $\forall n \in \mathbf{N}$, $a_n > 0$ and $a_{n+1} < a_n$ so we may use the condensation test.

$$2^n a_{2^n} = 2^n \cdot (1/2^n)^\alpha = 2^n \cdot 2^{-\alpha n} = (2^{1-\alpha})^n.$$

Therefore $\sum 2^n a_{2^n}$ is a geometric series of common ratio $2^{1-\alpha}$, and this converges if and only if $|2^{1-\alpha}| < 1$, that is, if and only if $1 - \alpha < 0$, which is equivalent to $\alpha > 1$. Therefore by the condensation test, $\sum 1/n^\alpha$ converges if and only if $\alpha > 1$.

The series $\sum 1/n^\alpha$ are useful as yardsticks in the comparison test. To be truthful, the condensation test is useful for few other results, though some series involving powers and logarithms are susceptible to this attack.

Example Decide whether $\sum (n+1)/(n^3 + 2n)$ is convergent or divergent.

Solution: For large $n, (n+1)/(n^3 + 2n)$ ought to behave like $1/n^2$ so, since $\sum 1/n^2$ converges, we show the series converges by comparison. Let $a_n = (n+1)/(n^3 + 2n)$ and $b_n = 1/n^2$. $\sum b_n$ converges and $a_n/b_n \to 1$ as $n \to \infty$ so Corollary 3 to the comparison test shows that $\sum a_n$ converges.

Although we have established enough theory to be able to tackle a range of examples, this is not wide enough to give useful experience at this stage. This is partly because we have yet to develop routine techniques and partly because the experience of judging which test to try first is best gained when the main tests are available. So we shall march on and expand our repertoire to allow series with terms of varying sign.

Theorem 5.7   Let $\sum a_n$ be a series of real numbers (positive or negative). Then if $\sum |a_n|$ converges so does $\sum a_n$.

Proof: Let

$$b_n = \begin{cases} a_n & \text{if } a_n \geq 0, \\ 0 & \text{if } a_n < 0, \end{cases} \qquad c_n = \begin{cases} 0 & \text{if } a_n \geq 0, \\ -a_n & \text{if } a_n < 0. \end{cases}$$

Then $\forall n \in \mathbf{N}$ $b_n \geq 0$ and $c_n \geq 0$ and $a_n = b_n - c_n$. Also, $0 \leq b_n \leq |a_n|$ and $0 \leq c_n \leq |a_n|$, so if $\sum |a_n|$ is convergent the comparison test proves that $\sum b_n$ and $\sum c_n$ are convergent, and thus so is $\sum (b_n - c_n)$. $\square$

Definition   A series for which $\sum |a_n|$ is convergent is said to be **absolutely convergent**. We have just proved that an absolutely convergent series is convergent in the ordinary sense.

Corollary   If $\sum a_n$ is a series of real numbers, $\sum b_n$ is a convergent series of non-negative numbers and for some integer N we have $\forall n \geq N$ $|a_n| \leq b_n$, then $\sum a_n$ is (absolutely) convergent.

Proof: By the earlier forms of comparison test, $\sum |a_n|$ converges. $\square$

In practice we refer to the foregoing result together with the comparison test and all its Corollaries as 'the comparison test', without distinguishing which component we are referring to. The last form is the most useful, largely because the majority of important series are absolutely convergent. However, all forms of the comparison test suffer from the disadvantage that they require the user to invent a series with which to compare the given series. We now develop some more routine tests, the main two, the ratio and root tests, being based on comparison with a geometric series.

Theorem 5.8 (The ratio test)   Let $\sum a_n$ be a series of real numbers for which $|a_{n+1}/a_n| \to L$ as $n \to \infty$.
Then if $L < 1$ the series converges (absolutely);
        if $L > 1$ the series diverges;
and if $L = 1$ we obtain no information.

Proof: Suppose first that $L < 1$. Let $\varepsilon = (1 - L)/2$ so that $\varepsilon > 0$ and $L + \varepsilon < 1$.
Then $\exists N$ s.t. $\forall n \geq N$ $|a_{n+1}/a_n| < L + \varepsilon$.

$$\therefore \exists N \quad \text{s.t.} \quad \forall n \geq N \quad |a_{n+1}| < (L + \varepsilon)|a_n|.$$
$$\therefore \exists N \quad \text{s.t.} \quad \forall n \geq N \quad |a_n| < (L + \varepsilon)^{n-N}|a_N|.$$

Since $|a_N|$ and $L + \varepsilon$ are independent of n and $\sum (L + \varepsilon)^n$ converges (for $0 < L + \varepsilon < 1$), the comparison test now shows that $\sum |a_n|$ converges.

Next, let $L > 1$. Then let $\varepsilon = L - 1$ so $\varepsilon > 0$ and $\exists N$ s.t. $\forall n \geq N$ $|a_{n+1}/a_n| > L - \varepsilon = 1$.

$$\therefore \exists N \quad \text{s.t.} \quad \forall n \geq N \quad |a_{n+1}| > |a_n|.$$
$$\therefore \forall n \geq N \quad |a_n| \geq |a_N|.$$

Since $|a_N| \neq 0$ (or else a_{N+1}/a_N would not exist) the last line shows that $a_n \not\to 0$, so by our first result on series, $\sum a_n$ diverges.

To see that the case $L = 1$ really gives no information, consider $\sum 1/n$ and $\sum 1/n^2$. In both cases, $|a_{n+1}/a_n| \to 1$ as $n \to \infty$, but one of the series converges and the other diverges. □

Example For all real x, $\sum x^n/n!$ converges.

Solution: If $x = 0$ the result is obvious, so suppose $x \neq 0$ and let $a_n = x^n/n!$ ($\neq 0$). Then $|a_{n+1}/a_n| = |x|/(n+1) \to 0$ as $n \to \infty$ and since $0 < 1$ the series converges by the ratio test.

Theorem 5.9 (The root test) Suppose that $\sum a_n$ is a series of real numbers and that $|a_n|^{1/n} \to L$ as $n \to \infty$.
Then if $L < 1$ the series converges (absolutely);
       if $L > 1$ the series diverges;
and if $L = 1$ we obtain no information.

Proof: Let $L < 1$. Set $\varepsilon = (1 - L)/2 > 0$ so $\exists N$ s.t. $\forall n \geq N$ $|a_n|^{1/n} < L + \varepsilon$ whence $|a_n| < (L + \varepsilon)^n$. Since $L + \varepsilon < 1$, the series $\sum (L + \varepsilon)^n$ converges, so $\sum |a_n|$ converges by comparison.

Let $L > 1$. This time set $\varepsilon = L - 1 > 0$ so $\exists N$ s.t. $\forall n \geq N$ $|a_n|^{1/n} > L - \varepsilon = 1$ whence $\forall n \geq N$ $|a_n| > 1$. Thus $a_n \not\to 0$ and $\sum a_n$ diverges.

As with the ratio test, the two series $\sum 1/n$ and $\sum 1/n^2$ show that the case $L = 1$ can occur with a convergent or a divergent series. □

It is worth noticing that both these tests, in the form in which we have given them, only give information if the terms of $\sum a_n$ are sufficiently regular that the limit exists. In practice this covers most of the cases which arise but notice that the proof of the ratio test would show that $\sum a_n$ is absolutely convergent if there is a constant $k < 1$ and $N \in \mathbf{N}$ s.t. $\forall n \geq N$ $|a_{n+1}/a_n| \leq k$, since the first step of the proof is devoted to using the limit condition to establish this (with $L + \varepsilon$ in place of k). It is, however, easy to mistake this result, since k must be independent of n and less than 1; for various plausible but wrong results, see the problems at the end of the chapter. The root test will, in principle, give information in every case which the ratio test copes with and some additional ones and is thus theoretically better, though in practice the nth roots are more awkward to handle. Again, the details are in the problems.

All of the tests so far will only indicate that a series converges absolutely, and are not capable of distinguishing a series which is convergent but not absolutely convergent. The only test we shall give for this is:

Theorem 5.10 (The alternating series theorem) Suppose that (a_n) is a decreasing sequence of positive numbers and that $a_n \to 0$ as $n \to \infty$. Then $\sum (-1)^{n-1} a_n$ converges.

Proof: Let s_n denote the sum of the first n terms of $\sum (-1)^{n-1} a_n$, so

$s_n = a_1 - a_2 + a_3 - a_4 + \cdots + (-1)^{n-1} a_n$. For $m \in \mathbf{N}$ and $n \geq 2m$,

$$s_n = s_{2m} + a_{2m+1} - a_{2m+2} + \cdots + (-1)^{n-1} a_n$$

$$= \begin{cases} s_{2m} + (a_{2m+1} - a_{2m+2}) + \cdots + (a_{n-1} - a_n) & (n \text{ even}), \\ s_{2m} + (a_{2m+1} - a_{2m+2}) + \cdots + (a_{n-2} - a_{n-1}) + a_n & (n \text{ odd}). \end{cases}$$

Since each bracket is non-negative and $a_n \geq 0$ we see that $s_n \geq s_{2m}$. Similarly if $n \geq 2m+1$

$$s_n = \begin{cases} s_{2m+1} - (a_{2m+2} - a_{2m+3}) - \cdots - (a_{n-2} - a_{n-1}) - a_n & (n \text{ even}), \\ s_{2m+1} - (a_{2m+2} - a_{2m+3}) - \cdots - (a_{n-1} - a_n) & (n \text{ odd}), \end{cases}$$

so, since the brackets are non-negative, $s_n \leq s_{2m+1}$. Therefore

$$\forall n \geq 2m \quad s_{2m} \leq s_n \leq s_{2m+1}. \tag{$*$}$$

This shows that (s_n) is bounded, but we still need care. The sequence $(s_2, s_4, s_6, \ldots) = (s_{2n})$ is increasing, since $\forall m \in \mathbf{N}$ $s_{2m} \leq s_{2m+2}$, and (s_{2n}) is bounded above, by s_3 for example, hence (s_{2n}) tends to a limit as $n \to \infty$. Let $s_{2n} \to s$ as $n \to \infty$. Also $(s_1, s_3, s_5, \ldots) = (s_{2n+1})$ is decreasing, since $\forall n \in \mathbf{N}$ $s_{2(m+1)+1} \leq s_{2m+1}$, and bounded below by s_2 so (s_{2n+1}) also tends to a limit as $n \to \infty$. Call its limit t. Then $t - s = \lim_{n \to \infty} (s_{2n+1} - s_{2n}) = \lim_{n \to \infty} a_{2n+1} = 0$, so $s = t$.

Finally, $s_n \to s$ as $n \to \infty$. To see this, let $\varepsilon > 0$. Then:

$$\exists N_1 \quad \text{s.t.} \quad \forall m \geq N_1 \quad |s_{2m} - s| < \varepsilon \quad (s_{2m} \to s)$$

and

$$\exists N_2 \quad \text{s.t.} \quad \forall m \geq N_2 \quad |s_{2m+1} - s| < \varepsilon \quad (s_{2m+1} \to s).$$

Let $N = \max(2N_1, 2N_2 + 1)$. Then if $n \geq N$, either $n = 2m$ or $n = 2m+1$, where m is an integer; in the first case $m \geq N_1$, in the second $m \geq N_2$ so in both cases $|s_n - s| < \varepsilon$. Therefore $\forall n \geq N \, |s_n - s| < \varepsilon$.

Since $\varepsilon > 0$ was arbitrary we have proved that (s_n) tends to a limit as $n \to \infty$, that is, that $\sum (-1)^{n-1} a_n$ is convergent. $\quad \square$

Notice the inequality $(*)$. Keeping m fixed and letting $n \to \infty$ allows us to deduce that $s_{2m} \leq s \leq s_{2m+1}$ giving a simple estimate for $\sum_{n=1}^{\infty} (-1)^{n-1} a_n$.

Example By the alternating series theorem, the series $1 - \dfrac{1}{2} + \dfrac{1}{3} - \dfrac{1}{4} + \cdots$, that is, $\sum (-1)^{n-1}/n$, converges. Since $\sum 1/n$ diverges, we have discovered a series which converges but is not absolutely convergent.

Infinite series are not just sums of infinite collections of numbers, or, at least, that is not how the sum to infinity was defined. We must, therefore, be careful not to presume without proof that all of the properties of finite sums hold for sums of series. In fact, although the expected properties do hold under suitable conditions, some may fail to be true in complete generality.

Examples Consider the series $\sum (-1)^{n-1}/n$. Let $s_n = 1 - \dfrac{1}{2} + \dfrac{1}{3} + \cdots +$

$(-1)^{n-1}/n$ be the sum of the first n terms. We know that for some $s, s_n \to s$ as $n \to \infty$ and the estimate in the proof of the alternating series theorem allows us to deduce that $s_2 \le s \le s_3$, that is, $\dfrac{1}{2} \le s \le \dfrac{5}{6}$. In particular, $s > 0$.

Now rearrange the order of the terms so that we add in turn the first positive term, the first two negative ones, the next positive term, the next two negative ones, ... The resulting series is

$$1 - \frac{1}{2} - \frac{1}{4} + \frac{1}{3} - \frac{1}{6} - \frac{1}{8} + \frac{1}{5} - \cdots,$$

and if we let t_n denote the sum of the first n terms of this series, then for $n \in \mathbf{N}$

$$t_{3n} = 1 - \frac{1}{2} - \frac{1}{4} + \frac{1}{3} - \cdots + 1/(2n-1) - 1/(4n-2) - 1/4n$$

$$= s_{2n} - (1/(2n+2) + 1/(2n+4) + \cdots + 1/(4n))$$

$$= s_{2n} - \left(\frac{1}{2}\right)(1/(n+1) + 1/(n+2) + \cdots + 1/(2n)).$$

Let $u_n = 1 + \dfrac{1}{2} + \dfrac{1}{3} + \cdots + \dfrac{1}{n}$, so that

$$1/(n+1) + 1/(n+2) + \cdots + 1/(2n) = u_{2n} - u_n = u_{2n} - 2\cdot(u_n/2)$$

$$= \left(1 + \frac{1}{2} + \cdots + \frac{1}{2n}\right) - 2\left(\frac{1}{2} + \frac{1}{4} + \cdots + \frac{1}{2n}\right) = s_{2n}.$$

Therefore $t_{3n} = s_{2n} - \frac{1}{2}s_{2n} = \frac{1}{2}s_{2n}$, so $t_{3n} \to \frac{1}{2}s \ne s$ as $n \to \infty$. By considering t_{3n+1} and t_{3n+2} it is not hard to show that $t_n \to \frac{1}{2}s$ as $n \to \infty$, that is, *the rearranged series converges to a different sum from the original one.*

A simpler point, easily missed, concerns brackets. The series $1 - 1 + 1 - 1 + \cdots$ diverges, but if we insert brackets to produce $(1-1) + (1-1) + (1-1) + \cdots$ all of the terms in the new series are 0, so *inserting or removing brackets may alter convergence.* If s_n denotes the sum of the first n terms of the original series then $s_n = 1$ or 0 according as n is odd or even. By putting in the brackets we obtain a series, the sum of whose first n terms is s_{2n} and $s_{2n} \to 0$ as $n \to \infty$.

These pathologies should not be a source of dismay. The seventeenth century view of series as an 'infinite sum' would regard these examples as paradoxes, since the less rigorous demonstrations of mathematical results in use then (and in much of school mathematics) tend to contradict them. Our more precise logical approach is capable of coping with these subtleties, and one by-product is these pathological results. These should not be given too much emphasis, except to counsel caution. In a sense, these apparently odd results indicate that our intuitive idea of an 'infinite sum' corresponds to something better behaved than the mere convergence of a series. The intuitive idea is closer to absolute convergence; though we shall not prove it here, a rearrangement of an *absolutely* convergent series converges to the same sum as the original series.

Examples   Discuss the convergence of $\sum nx^n$ and $\sum(x^{3n}/2^n n)$ for all real values of x.

Solution: The easiest tests are the ratio and root tests, so we try these first. Since $|(n+1)x^{n+1}/(nx^n)| = \{(n+1)/n\}|x| \to |x|$ as $n \to \infty$, the ratio test shows that the first series is convergent for $|x| < 1$ and divergent for $|x| > 1$. This leaves the case $|x| = 1$ undecided. However, since $|x| = 1$ implies $|nx^n| = n \not\to 0$ as $n \to \infty$ we see that the series diverges for $x = \pm 1$.

Applying the ratio test to the second series, we notice that if $a_n = x^{3n}/(2^n n)$ then (for $x \neq 0$) $|a_{n+1}/a_n| = (n/(n+1))(|x|^3/2)) \to |x|^3/2$ as $n \to \infty$. By the ratio test, then, the series converges if $|x|^3 < 2$ (including the trivial case when $x = 0$) and diverges if $|x|^3 > 2$. The ratio (and root) tests fail us if $|x| = 2^{1/3}$, so we proceed separately. Let $x = 2^{1/3}$; then $a_n = 1/n$ and the series is $\sum 1/n$ which we know diverges. Also, if $x = -2^{1/3}$, $a_n = (-1)^n/n$, which gives the series $\sum(-1)^n/n$, convergent by the alternating series theorem. (Strictly, the theorem shows $\sum(-1)^{n-1}/n$ converges, but multiplying each term by -1 does not affect the convergence.) In this case the series converges for $-2^{1/3} \leq x < 2^{1/3}$ and diverges for other real x.

These examples are fairly typical in that the ratio or root tests may be brought to bear on most cases (or most values of a parameter) and are thus worth trying first, before the others are tried. Experience will bear out this rule of thumb and show the type of example for which the ratio test yields no information.

Definition   A series of the form $\sum a_n x^n$ where, for each n, the number a_n does not depend on x, is called a **power series**. It is usual to allow the index of summation of a power series to start at $n = 0$ giving a term $a_0 x^0$ which we interpret as a_0 and which is independent of x.

Lemma 5.11   Suppose that the power series $\sum a_n w^n$ converges. Then the series $\sum a_n x^n$ converges absolutely for all x with $|x| < |w|$.

Proof: Since $\sum a_n w^n$ converges, $a_n w^n \to 0$ as $n \to \infty$ and so $(a_n w^n)$ is a bounded sequence. Choose M such that $\forall n \geq 0 \ |a_n w^n| \leq M$. Then for $|x| < |w|$ and $n \geq 0$, $|a_n x^n| = |a_n w^n| \cdot |x/w|^n \leq M|x/w|^n$. Since $|x/w|$ is independent of n and $|x/w| < 1$, $\sum|x/w|^n$ converges and so, by comparison, $\sum|a_n x^n|$ converges.   □

Corollary   Suppose $\sum a_n w^n$ diverges. Then $\sum a_n x^n$ diverges for all x with $|x| > |w|$.

Proof: If $\sum a_n x^n$ converged and $|x| > |w|$, then by the Theorem, $\sum a_n w^n$ would converge.   □

Lemma 5.11 and its Corollary tell us that a power series $\sum a_n x^n$ will converge if $|x|$ is 'small enough' and diverge if $|x|$ is 'large enough'. This may be put more precisely.

Definition   Let $\sum a_n x^n$ be a power series. We say that the number $R \geq 0$ is the **radius of convergence** of the series if $\sum a_n x^n$ converges for all x with $|x| < R$ and diverges for all x with $|x| > R$.

Considering a few of the examples that have arisen (or will arise when you do the problems!) we see that $\sum x^n/n$ converges if $|x| < 1$, diverges if $|x| > 1$, while for $|x| = 1$ the series converges for $x = -1$ and diverges for $x = +1$. Here the radius of convergence is 1, but notice that this fact on its own does not indicate the behaviour for $|x| = 1$. $\sum n! x^n$ has radius of convergence 0 (the ratio test shows that the series diverges if $|x| \neq 0$) while $\sum x^n/n!$ does not have a radius of convergence; it converges for all real x. We shall prove later that a power series either converges for all real x or has a radius of convergence. The examples given are typical of many, in that the ratio test will determine the convergence for all x except for $|x|$ equal to the radius itself where other tests must be involved.

Example   Let $\sum a_n x^n$ be a power series, and suppose that $|a_{n+1}/a_n| \to L$ as $n \to \infty$. Then if $L > 0$ the series has radius of convergence $1/L$ while if $L = 0$ the series converges for all real x. If $|a_{n+1}/a_n| \to \infty$ as $n \to \infty$ the series converges only for $x = 0$. The proof is a straightforward use of the ratio test.

Decimals

The traditional way of writing a number in decimal notation is in the form $b_N b_{N-1}\ldots b_0 \cdot a_1 a_2 a_3 \ldots$ where each of the symbols a_n, b_n denotes one of the digits $0, 1, \ldots, 9$. This represents the number $\sum_{n=0}^{N} b_n 10^n + \sum_{n=1}^{\infty} a_n 10^{-n}$, the first sum being an integer, the second the 'fractional part'. The second sum is an infinite series, and comparison with $\sum 9.10^{-n}$ shows that it converges. Therefore, every decimal expansion corresponds to a real number. The converse result, that every real number has a decimal expansion, is longer and there are numbers which can be expanded as decimals in more than one way. These assertions are proved in Appendix B.

Problems

1. Prove that $\sum_{n=1}^{N} 1/(n(n+1)(n+2)) = 1/4 - 1/(2(N+1)(N+2))$ and deduce that $\sum_{n=1}^{\infty} 1/(n(n+1)(n+2)) = 1/4$.

2. Let $a_n = 1/2^n$ if n is odd and $a_n = (3/4)^n$ if n is even. Show that $\sum a_n$ converges.

3. By comparing the following series with $\sum 1/n^\alpha$ for suitable α, decide whether each converges or diverges: $\sum (n+1)/(n^2+1)$, $\sum (n^2+1)/(n+1)$, $\sum (n+1)/(n^3+1)$, $\sum (n+1)/n^4$.

4. Suppose that $\forall n \in \mathbf{N}\ a_n \geq 0$ and that $\sum a_n$ converges. Prove that $\sum a_n^2$ converges. Give an example where $\sum b_n^2$ converges but $\sum b_n$ does not.

5. Given that $\sum a_n$ is absolutely convergent, prove that $|\sum_{n=1}^{\infty} a_n| \leq \sum_{n=1}^{\infty} |a_n|$.

6. Decide which of the following series converge: $\sum n/2^n$, $\sum 2^n/(n3^n)$,

$\sum(n2^n/3^n)$, $\sum 1/n^n$, $\sum(n+1)^2/2^n$, $\sum 2^n/n!$, $\sum(2n)!/(n!)^2$, $\sum(2n)!/(8^n n! n!)$, $\sum(-1)^{n-1}/\sqrt{n}$.

7. For which positive values of α does the series $\sum(-1)^{n-1}/n^\alpha$ converge?

8. Find for which real x the series $\sum x^n/(1+x^{2n})$ converges; make sure you test all x.

9. Find the radius of convergence of the following series, or, if there is none, show that the series converges for all real x: $\sum x^n/n^2$, $\sum 2^n x^n$, $\sum x^n/(n^2+1)$, $\sum x^n/n$, $\sum x^{2n}/n$, $\sum x^{2n}/2^n$, $\sum x^n/(n2^n)$, $\sum(1+1/n)^n x^n$, $\sum n! x^n$, $\sum n! x^n/(2n)!$, $\sum((2n)!/(n!n!))x^n$, $\sum((3n)!/(n!(2n)!))x^n$, $\sum(x/n)^n$.

10. In all but the last three parts of Q9, consider those x for which $|x|$ equals the radius of convergence and decide whether or not the series converges for those x.

11. By observing that if $n \geq k$, $n! \geq (k+1)^{n-k}k!$, show that $\sum_{n=k}^{\infty} 1/n! \leq (k+1)/(k \cdot k!)$. Let e denote the number $1 + \sum_{n=1}^{\infty} 1/n!$, and show that if $s_m = 1 + \sum_{n=1}^{m-1} 1/n!$, then $s_m \leq e \leq s_m + (m+1)/(m \cdot m!)$. Deduce that $2 \leq e \leq 3$.

12. Choose a sequence (a_n) which is strictly decreasing and for which $a_n \to 1$ as $n \to \infty$. Notice that for this sequence $\forall n \in \mathbb{N}$ $|a_{n+1}/a_n| < 1$ but $\sum a_n$ diverges since (a_n) does not even tend to 0.

13. Suppose that $|a_{n+1}/a_n| \to L$ as $n \to \infty$. Given $\varepsilon > 0$ show that for some N, $n \geq N \Rightarrow (L-\varepsilon)^{n-N}|a_N| < |a_n| < (L+\varepsilon)^{n-N}|a_N|$ and deduce that $|a_n|^{1/n} \to L$ as $n \to \infty$. Notice that this result shows that the root test gives a result whenever the ratio test does and is thus 'stronger'. In practice, the roots may be awkward to evaluate, and this result may help—for example, it may be used to show $(1/n!)^{1/n} \to 0$ as $n \to \infty$.

14. By showing that $\sum n^\alpha x^n$ converges, or otherwise, show that if α is fixed and $|x| < 1$, $n^\alpha x^n \to 0$ as $n \to \infty$.

15. Suppose that $\sum a_n$ and $\sum b_n$ are both convergent and that $\forall n \in \mathbb{N}$ $a_n \leq b_n$. Show that $\sum_{n=1}^{\infty} a_n \leq \sum_{n=1}^{\infty} b_n$. Given the additional information that there is an $N \in \mathbb{N}$ for which $a_N < b_N$, show that $\sum_{n=1}^{\infty} a_n < \sum_{n=1}^{\infty} b_n$. [Hint: Let (s_n) and (t_n) be the corresponding sequences of partial sums. It is *not* enough merely to notice that $\forall n \geq N$ $s_n < t_n$ since this only guarantees that $\lim s_n \leq \lim t_n$. Notice that $\forall n \geq N$ $s_n \leq t_n - (b_N - a_N)$ where $b_N - a_N$ is positive and independent of n.]

16. Suppose that $\sum a_n$ is convergent and $\sum b_n$ is divergent. Prove that $\sum(a_n + b_n)$ is divergent.

17. Suppose that $\sum a_n$ is convergent, but not absolutely convergent, and let $b_n = \max(a_n, 0)$ and $c_n = \min(a_n, 0)$, so that $a_n = b_n + c_n$ and $|a_n| = b_n - c_n$. Prove that $\sum b_n$ and $\sum a_n$ both diverge.

18.* Let (a_n) be a sequence such that $a_n = 0$ if n is odd and let $b_n = a_{2n}$. Show that $\sum b_n$ converges if and only if $\sum a_n$ does and that $\sum_{n=1}^{\infty} b_n = \sum_{n=1}^{\infty} a_n$. Can you generalise this?

19.* Let $\sum a_n$ and $\sum b_n$ be convergent series and define (c_n) by $c_{2n} = b_n$ $c_{2n-1} = a_n$, so (c_n) is $(a_1, b_1, a_2, b_2, \ldots)$. Prove that $\sum c_n$ converges and $\sum_{n=1}^{\infty} c_n = \sum_{n=1}^{\infty} a_n + \sum_{n=1}^{\infty} b_n$.

20.* Using the notation of Q11 show that e is irrational by observing that if p and q are positive integers, $|p/q - s_{q+1}| \leq (q+2)/((q+1)(q+1)!)$ and $q!|p/q - s_{q+1}|$ is an integer, if $e = p/q$. Obtain a contradiction.

21. Show that for integers k, n satisfying $0 \leq k \leq n$, $0 \leq \binom{n}{k} 1/n^k \leq 1/k!$

and hence that $\forall n \in \mathbf{N}$ $2 \leq (1 + 1/n)^n \leq e$. Notice, in particular, that, however the sequence behaves, it does not tend either to 1 (which is $\lim_{n \to \infty} (1 + 1/n)^k$) or to ∞ (which is $\lim_{n \to \infty} (1 + 1/k)^n$) where k here denotes a fixed natural number.

CHAPTER 6

The Structure of the Real Number System

"The integers God made: all the rest is the
work of man."

L. Kronecker

The invention of the calculus presented the mathematical world with a severe
difficulty. No one was in doubt of its usefulness, nor that it could produce
many correct results that could not be obtained by other methods, yet the
foundations of the calculus were unsatisfactory. There was some doubt as to
the exact nature of a limit, and proofs of various results left much to be
desired—indeed, even from the earliest times of the calculus, mathematicians
had produced 'proofs' of results which were visibly false. The calculus became
a matter of some controversy, well exemplified in Berkeley's book, *The
Analyst, or A Discourse Addressed to an Infidel Mathematician*, published in
1734; this is still worth reading.

As is human nature, mathematicians, when faced with the apparently
insoluble problem of putting calculus on a rigorous foundation, tried to
ignore the matter, and until the end of the eighteenth century calculus was
accepted largely as a matter of faith. By that time, however, the issue had
become pressing, and major progress was made in the early nineteenth
century, with the rigorous development of the idea of limit and its consequences. The stumbling block at last became apparent; there was no precise
statement of the properties of the real number system. This was rectified
around 1870 and the modern subject of analysis assumed increasing
importance thereafter.

To an extent, we have followed a similar path. We introduced the idea of
limit and deduced its consequences, but we noticed that there were results
we needed to assume. For example, we did not prove the theorem that a
bounded increasing sequence must tend to some real limit, but deferred this
issue. The important thing is to appreciate that there *is* an issue here; no
matter how plausible the statement may seem, it does require justification
unless it is to be accepted as a basic premiss. If we accept the property as
true, for the moment, then it provides a distinction between the set of real
numbers and the set of rational numbers. To see this, notice that if
$s_n = 1 + 1/1! + 1/2! + \cdots + 1/n!$, then (s_n) is increasing and bounded above by
3 but the limit, the number usually called e, is not rational even though all

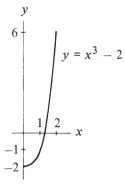

Fig. 6.1

of the terms s_n are rational. (These assertions are the substance of Problems 5.11 and 5.20.)

To cope with this task we need to add to our axioms for **R**, since the result we wish (that an increasing sequence of real numbers which is bounded above tends to some real limit) cannot be proved from the axioms we have so far adopted. The axioms A1–A12 assumed at the beginning of Chapter 3 are true of the system of rational numbers, so any result which can be deduced from these axioms alone must be true of the system of rational numbers and, indeed, of any other system satisfying A1–A12. Since we know that a bounded increasing sequence in the system of rational numbers need not have a limit within that system, the property we wish is not true of **Q** and therefore not deducible from A1–A12 alone.

We have already stated that **R** is a set with various arithmetical properties and our picture of the real number system is that it corresponds to the points of a line. We therefore wish a property which would allow us to replace the following with something more precise (see Fig. 6.1): The function $y = x^3 - 2$ has a negative value at $x = 1$ and a positive one at $x = 2$, so, since the curve has no discontinuities, it 'must' cross the axis at some point. This is not so much a proof as a confession of failure of imagination ('I can't see how else y could change from being negative to being positive').

The feature we shall make precise is the idea that there are no gaps in the set of real numbers, as we think of there being no gaps in a line. (There certainly are gaps in the set of rational numbers and $\sqrt[3]{2}$ lies in one of them.) Looking more closely, we see that we can isolate two subsets of **R**, $\{x \in \mathbf{R} : x^3 - 2 < 0\}$ and $\{x \in \mathbf{R} : x^3 - 2 > 0\}$, with the property that every member of the first set is less than every member of the second set. The existence of the number $\sqrt[3]{2}$ is equivalent to the existence of a real number which marks the division between these two sets. This corresponds to a picture of dividing a line into two: if we divide the line into two parts, one lying entirely to the left of the other, then there must be a point which marks the division, and this point must lie in one portion or the other (see Fig. 6.2). This is what motivates the definition below. (Recall that the union of two sets A and B is $A \cup B = \{x : x \in A \text{ or } x \in B\}$.)

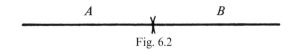

Fig. 6.2

Axiom of Continuity   (Dedekind's Axiom)

The set **R** has the property that if $\mathbf{R} = A \cup B$ where A and B are both non-empty and $\forall x \in A$, $\forall y \in B$ $x < y$, then either A has a greatest element or B has a smallest.

   Notice that the conditions ensure that A and B have no elements in common, and that the axiom guarantees that certain numbers exist.

Notation: Let $a, b \in \mathbf{R}$ and $a \le b$. We define the **closed interval** $[a, b]$ to be the set $\{x \in \mathbf{R} : a \le x \le b\}$ and the **open interval** (a, b) to be $\{x \in \mathbf{R} : a < x < b\}$. $(a, b]$ and $[a, b)$ have their obvious meaning. For convenience we use a similar notation for $\{x \in \mathbf{R} : x \le a\}$, denoting this by $(-\infty, a]$ and $\{x \in \mathbf{R} : x > a\}$ by (a, ∞), with the corresponding interpretation of $(-\infty, a)$ and $[a, \infty)$. As before, ∞ is used as a notation, not a number.

Theorem 6.1   Let $\mathbf{R} = A \cup B$ where A and B are non-empty and $\forall x \in A$, $\forall y \in B$ $x < y$. Then there is a real number a such that *either* $A = (-\infty, a]$ and $B = (a, \infty)$ *or* $A = (-\infty, a)$ and $B = [a, \infty)$. $\{x \in \mathbb{R}, x \le a]$

$B = \{x \in \mathbb{R} \cdot x > a]$ $A = \{x \in \mathbb{R} : x < a\}$ $B = \{x \in \mathbb{R} \cdot x > a\}$

Proof: By the axiom of continuity, either A has a greatest element or B has a smallest.

   First, suppose A has a greatest element, a. Then $a \in A$ and $\forall x \in A$ $x \le a$ so A contains only numbers belonging to $(-\infty, a]$; we need to show A contains all such numbers. Let $y \in \mathbf{R}$, $y \le a$. Since $y \in \mathbf{R}$, $y \in A$ or $y \in B$. But $y \in B \Rightarrow a < y$ (since $a \in A$, by the condition given connecting A and B), so this is impossible. Thus $y \in A$ so $A = (-\infty, a]$, since y was a typical number in $(-\infty, a]$. Since $B = \{x \in \mathbf{R} : x \notin A\} = \{x \in \mathbf{R} : x \not\le a\}$ $B = (a, \infty)$. (That $B = \{x \in \mathbf{R} : x \notin A\}$ uses our knowledge that $A \cap B = \varnothing$.)

   The case when B has a smallest element is similar. $\square$

   In practice, the Axiom of Continuity is inconvenient to use, so we must develop a more readily applicable tool.

Definitions   A subset S of **R** is said to be **bounded above** if there is a number u with the property that $\forall x \in S$ $x \le u$. Such a number u is called an **upper bound** for S. Notice that if u is an upper bound for S so are $u + 1, u + 2, \ldots$ so a set which is bounded above will have many upper bounds.

Theorem 6.2   Let S be a non-empty subset of **R** which is bounded above. Then among all of the upper bounds of S there is a smallest upper bound, i.e. the set of upper bounds for S has a smallest element.

Proof: Let B be the set of all upper bounds for S; we know $B \ne \varnothing$. Let A be the set of all real numbers which are not upper bounds for S. From the definitions it is obvious that $A \cup B = \mathbf{R}$. Since $S \neq \varnothing$, there is an element $s \in S$, whence $s - 1$ is not an upper bound for S and $A \neq \varnothing$.

To apply the Axiom of Continuity we need to show that every element of A is less than every element of B. Let $x \in A$ and $y \in B$. Since x is not an upper bound for S, there is an element, say s, of S s.t. $x < s$. Since y is an upper bound for S, $s \leq y$. Thus $x < y$. Because x and y were typical elements of A and B, we have shown that $\forall x \in A$, $\forall y \in B$ $x < y$.

By the Axiom of Continuity, either A has a largest element or B has a smallest. In fact, A cannot have a largest element. Let $a \in A$. Then a is not an upper bound for S, so there is an element of S, say s, such that $a < s$. Then also $\frac{1}{2}(a + s) < s$ so $\frac{1}{2}(a + s)$ is not an upper bound for S, whence $\frac{1}{2}(a + s) \in A$ and a is not the greatest element of A. Since a was a typical element of A, A can have no greatest element and hence B has a smallest. This is what we require. $\square$

Comments: 1. This result does tell us something significant. The set $(0, 1)$ has no smallest element even though all its members are greater than 0.
2. We do not claim that S has a largest element, but if it does, this element is the smallest upper bound. (For it is an upper bound and no smaller number is.)

Example   The smallest upper bound for $[0, 1]$ is 1, and the smallest upper bound for $(0, 1)$ is also 1. Notice that 1 belongs to one of the sets but not the other.

Definition   Let S be a non-empty set which is bounded above. We call the smallest upper bound for S the **supremum** of S and denote it by $\sup S$.

Lemma 6.3   Let S be a subset of $\mathbf{R}$. Then $s = \sup S$ if and only if (i) $\forall x \in S$ $x \leq s$ and (ii) $\forall s' < s$ $\exists x \in S$ s.t. $s' < x$.

Proof: Statement (i) says that s is an upper bound for S and (ii) that no smaller number is an upper bound. A moment's thought will show that a number possesses both properties if and only if it is the smallest upper bound for S. $\square$

The two conditions of Lemma 6.3 offer a practical method for establishing that a particular number is the supremum of a given set. Notice that there are two conditions to be checked and that this complication is intrinsic; the supremum need not be a member of the set concerned. Matters would be greatly simplified if every set with an upper bound had a greatest element, but it just is not true. Let us use the idea to prove one of our deferred results.

Theorem 6.4   Let (a_n) be an increasing sequence of real numbers which is bounded above. Then there is a real number a such that $a_n \to a$ as $n \to \infty$.

Proof: We know that $\{a_n : n \in \mathbf{N}\}$ is bounded above; it is non-empty since it contains a_1. Let $a = \sup\{a_n : n \in \mathbf{N}\}$.

Let $\varepsilon > 0$. Since $a - \varepsilon < a$, by Lemma 6.3 there is an element of $\{a_n:n\in\mathbf{N}\}$ which is greater than $a - \varepsilon$; let this element be a_N. Then, since (a_n) is increasing we see that $\forall n \geq N\ a - \varepsilon < a_N \leq a_n \leq a$, the last inequality being true since $a = \sup\{a_n:n\in\mathbf{N}\}$. Therefore $\forall n \geq N\ |a_n - a| < \varepsilon$. Since $\varepsilon > 0$ was not further specified, this holds for all $\varepsilon > 0$ and we have shown $a_n \to a$ as $n \to \infty$. $\square$

Notice that Lemma 6.3 allows us to deduce the existence of at least one element of the set $\{a_n:n\in\mathbf{N}\}$ greater than $a - \varepsilon$; we do not know whether or not $a - \varepsilon$ itself belongs to the set.

This leaves us in a position to tackle the detailed structure of $\mathbf{R}$. We base all our deductions on the assumptions we have made about $\mathbf{R}$, which are that $\mathbf{R}$ satisfies the arithmetical and order axioms A1–A12 stated at the beginning of Chapter 3 and that $\mathbf{R}$ obeys the axiom of continuity. By the arithmetical properties, we can prove that $\mathbf{R}$ contains rational numbers (e.g. $2/3 = (1 + 1)\cdot(1 + 1 + 1)^{-1}$, whose construction uses the arithmetical properties). To show that $\mathbf{R}$ contains irrational numbers we need the Axiom of Continuity, which we use indirectly. Define a sequence (a_n) by $a_1 = 2$, $a_{n+1} = \frac{1}{2}(a_n + 2/a_n)$. Then, following work before, we see that (a_n) is decreasing and bounded below, so there is a real number a such that $a_n \to a$ as $n \to \infty$. (This uses Theorem 6.4 above, which we have now proved.) As before, $a^2 = 2$ so a is not rational. This guarantees the existence of a real number whose square is 2; the details are in the example after the Corollary to Theorem 4.4.

The rational numbers are distributed throughout $\mathbf{R}$ and are thoroughly intermingled with the irrational numbers. Before showing this we prove the Archimedean property which we previously had to assume.

Theorem 6.5 The set of natural numbers is not bounded above. In other words, for every real number x, there is a natural number n for which $n > x$.

Remarks: This result may seem obvious. It certainly is obvious that $\mathbf{N}$ has no greatest element, though that on its own does not preclude the existence of an upper bound. Theorem 6.5 can be viewed as showing that $\mathbf{R}$ contains no 'infinitely large' numbers and is a form of confirmation that there are no further additional properties which we shall have to assume. Though it need not concern us here, there do exist systems satisfying A1–A12 for which the set corresponding to $\mathbf{N}$ is bounded above.

Proof: Suppose $\mathbf{N}$ were bounded above. Then, since $\mathbf{N} \neq \varnothing$, there must be a supremum of $\mathbf{N}$; call it s. Then, by the properties of the supremum, we see that since $s - \frac{1}{2} < s$, there is a natural number n with $s - \frac{1}{2} < n$. Then $n + 1 > s + \frac{1}{2} > s$, which is a contradiction since $n + 1 \in \mathbf{N}$ but $n + 1 > \sup\mathbf{N}$. This contradiction shows $\mathbf{N}$ cannot be bounded above. The second part of the statement is just the observation that if $x\in\mathbf{R}$ then x is not an upper bound for $\mathbf{N}$. $\square$

Corollary If $x\in\mathbf{R}$ and $x > 0$ then there is a natural number n with $1/n < x$.

Proof: Since $x > 0$, $1/x > 0$ so by Theorem 6.5 $\exists n\in\mathbf{N}$ s.t. $n > 1/x$. For this n, $1/n < x$. $\square$

This result shows us that if x is positive, no matter how small, then if we add x to itself sufficiently often, the result is greater than 1; this just notices that $\exists n \in \mathbf{N}$ s.t. $nx > 1$ and $nx = x + x + \cdots + x$. This could be interpreted as showing that there are no 'infinitesimally small' elements of $\mathbf{R}$.

Lemma 6.6   If x and y are real numbers and $x < y$, there is a rational number q with $x < q < y$. (Loosely, between every two reals there is a rational number.)

Proof: Since $y - x > 0$ there is $N \in \mathbf{N}$ with $1/N < y - x$. We now show that for some integer m, m/N is between x and y.

Since Nx is real, there is an $m_1 \in \mathbf{N}$ for which $m_1 > Nx$ so that $x < m_1/N$. Since $-Nx \in \mathbf{R}$, there is an $m_0 \in \mathbf{N}$ for which $m_0 > -Nx$, hence $x > -m_0/N$. The set $A = \{m/N : m \in \mathbf{Z}, -m_0 \le m \le m_1\}$ is a finite set which contains an element greater than x and another less than x. If m_2/N is the smallest member of A which is greater than x then $(m_2 - 1)/N \le x$. (This number exists since every *finite* set has a minimum.) Let $q = m_2/N$ so $q \in \mathbf{Q}$, $x < q$ and $q - 1/N \le x$. Thus $q \le x + 1/N < x + (y - x)$ whence $x < q < y$.   □

Obviously, we can repeat the process just carried out to produce a rational number q_1 with $x < q_1 < q < y$, and repeat again to produce as many rationals as we wish between x and y. There are, therefore, infinitely many rationals between x and y, since if this were not so there would be some maximum number of rationals which could be found between x and y.

Corollary 1   If $x \in \mathbf{R}$ then $x = \sup \{q \in \mathbf{Q} : q < x\}$.

Proof: Let $A = \{q \in \mathbf{Q} : q < x\}$. We show the result by using Lemma 6.3. Obviously, $\forall y \in A$, $y \le x$. Also if $y' < x$ then by Lemma 6.6 there is a $q \in \mathbf{Q}$ with $y' < q < x$ so $\exists q \in A$ s.t. $y' < q$. By Lemma 6.3, $x = \sup A$.   □

Corollary 2   Every real number is the limit of a sequence of rational numbers.

Proof: Let $a \in \mathbf{R}$. For each $n \in \mathbf{N}$, there is a rational number a_n with $a - 1/n < a_n < a$, by the Lemma. Then $\forall n \in \mathbf{N}$ $|a_n - a| < 1/n$ so $a_n \to a$ as $n \to \infty$.   □

In the foregoing work we have introduced and used the idea of the supremum of a set. This is an important tool, so we need to obtain some skill in using it. Notice that the definition of the supremum as the smallest upper bound is somewhat complicated, and the equivalent statement in Lemma 6.3 has two parts. It is usual to have to establish these two properties of the supremum separately.

Example   Let A be a non-empty subset of $\mathbf{R}$ which is bounded above, and let $B = \{2x : x \in A\} = \{y : y/2 \in A\}$. Then $\sup B = 2 \sup A$.

Solution: Let $a = \sup A$. We need to show that $2a$ has the two properties required by Lemma 6.3.

Let $y \in B$. Then $y = 2x$ for some $x \in A$, so $y/2 \in A$. Thus $y/2 \le a$ (as $a = \sup A$), and $y \le 2a$. Therefore $\forall y \in B$, $y \le 2a$.

Let $b' < 2a$. Then $b'/2 < a$. Since $a = \sup A$, from the second part of Lemma 6.3, $\exists x \in A$ s.t. $b'/2 < x$ and so $b' < 2x$. Thus $\exists y \in B$ ($y = 2x$ before) s.t. $b' < y$. This establishes the condition $\forall b' < 2a \; \exists y \in B$ s.t. $b' < y$.

We have established both conditions of Lemma 6.3, so $2a = \sup B$.

The two-part nature of the proof here is typical.

Definitions A real number l is said to be a **lower bound** for the set A if $\forall x \in A \; x \geq l$. A set is **bounded below** if it possesses a lower bound. The real number m is called the **infimum** of the set A and denoted by $\inf A$ if (and only if) it is the greatest lower bound of A.

A set is said to be **bounded** if it is both bounded above and bounded below.

To save repeating our work to produce the properties of inf, we notice:

Lemma 6.7 Let A be a non-empty set of real numbers which is bounded above and let $B = \{x: -x \in A\}$. Then B is bounded below and $\inf B = -\sup A$.

Proof: Let $a = \sup A$. Then $x \in B \Rightarrow -x \in A \Rightarrow -x \leq a \Rightarrow x \geq -a$. Thus $-a$ is a lower bound for B. In passing, notice this shows B is bounded below.

Suppose that b is a lower bound for B; we need to relate this to A. $\forall x \in A$, $-x \in B$ so $-x \geq b$ and $x \leq -b$. Thus $\forall x \in A, x \leq -b$. Therefore $-b$ is an upper bound for A, hence $a \leq -b$, since $a = \sup A$ is the smallest upper bound for A. Hence $b \leq -a$. We have shown that if b is a lower bound for $B, b \leq -a$ so, since $-a$ is a lower bound for B, it is the greatest lower bound, that is, $\inf B$. □

Lemma 6.8 Every non-empty set of real numbers which is bounded below has a (real) infimum. If A is a subset of **R**, then $t = \inf A$ if and only if (i) $\forall x \in A \; x \geq t$ and (ii) $\forall t' > t \; \exists x \in A$ s.t. $t' > x$.

Proof: Let A be non-empty and bounded below. Let $C = \{c: -x \in A\}$. Then C is non-empty and bounded above and thus C has a supremum. (If l is a lower bound for A, $-l$ is an upper bound for C, as is easily checked.) By Lemma 6.7, then, $\inf A = -\sup C$ so $\inf A$ exists. The proof that the criteria are correct can either be obtained by relating them to $-t = \sup C$ or by analogy with the proof of Lemma 6.3. □

Lemma 6.9 If A is a bounded non-empty subset of **R**, $\inf A \leq \sup A$.

Proof: Since $A \neq \emptyset$, we can choose $x \in A$. Then $\inf A \leq x$ and $x \leq \sup A$ so $\inf A \leq \sup A$. □

Example Let A and B be two non-empty sets which are bounded above and let $C = \{x + y: x \in A$ and $y \in B\}$. Show that $\sup C = \sup A + \sup B$.

Solution: Let $a = \sup A$ and $b = \sup B$. Then if $z \in C$ there are $x \in A$ and $y \in B$ s.t. $z = x + y \leq a + b$ (since $x \leq a$, $y \leq b$). Thus $\forall z \in C \; z \leq a + b$.

Now let $c' < a + b$. Then $c' - b < a$ so, since $a = \sup A, \exists x \in A$ with $c' - b < x$.

For this value of x, $c' - x < b = \sup B$, so $\exists y \in B$ with $c' - x < y$. Thus, letting $z = x + y$, we have $z \in C$ s.t. $c' < z$. This gives the second property of $\sup C$, so $a + b = \sup C$ (by Lemma 6.3).

Example Suppose that A and B are two non-empty sets in **R** with the property that $\forall x \in A$ and $\forall y \in B$ $x \leq y$. Show $\sup A \leq \inf B$.

Solution: Let $x \in A$. Then $\forall y \in B$ $x \leq y$ so that x is a lower bound for B. Therefore $x \leq \inf B$. ($\inf B$ is the greatest lower bound.)

The conclusion of the above paragraph was reached only on knowing $x \in A$ so it must hold for all such x, hence $\forall x \in A$ $x \leq \inf B$. Thus $\inf B$ is an upper bound for A, so $\sup A \leq \inf B$.

Experience will show that in most examples where a supremum or infimum is to be shown equal to a given number, the proof is in two steps. Where an inequality involving sup or inf is concerned, one step may suffice and the two parts of Lemma 6.3 can be viewed as establishing two opposite inequalities (e.g. if a is an upper bound for A then $\sup A \leq a$).

Notice that the supremum of a set need not belong to the set (consider $(0, 1)$); life would be much simpler if this were true! Also notice that if $a' < \sup A$ $\exists x \in A$ s.t. $a' < x$. This guarantees the existence of some $x \in A$ with $a' < x$ but not that a' itself is in A. (Consider $A = \{1/2, 2/3, 3/4, \ldots\}$.)

We give closed intervals their own name because they have useful properties, one of which is that the closed interval $[a, b]$ contains both its supremum and infimum. An open interval contains neither its supremum nor its infimum. In the course of the next chapter we shall encounter several situations where the distinction is crucial.

There is one common type of set which is simpler in this respect. Subsets of **N**, other than the empty set, necessarily have a smallest element so in this case the set contains its infimum. This was proved in Lemma 3.3.

Problems

1. The following sets are bounded above; identify the set of upper bounds in each case: $(0, 1)$, $\{x: 1/x > 2\}$, $\{x \in \mathbf{Q}: x < 4\}$, $\{x \in \mathbf{Q}: x \leq 4\}$, $\{1/2, 2/3, 3/4, \ldots\}$, $\{1/2, 1/3, 1/4, \ldots\}$.

2. Let A be a non-empty subset of **R** such that $\forall x \in A$, $x < 2$. Prove that $\sup A \leq 2$. Give two examples of sets A satisfying this property, one with $\sup A = 2$, one with $\sup A < 2$.

3. Find the supremum of each of the sets in Q1 and check that the properties demanded by Lemma 6.3 are satisfied.

4. Show that if r is an irrational number and q is a non-zero rational number, rq is irrational. Use this to show that if x and y are real numbers and $x < y$ then there is an irrational number s with $x < s < y$. (Hint: Consider $x/\sqrt{2}$ and $y/\sqrt{2}$.)

5. Show that for each $n \in \mathbf{N}$ there is a rational number a_n satisfying

$\sqrt{2} - 1/n < a_n < \sqrt{2}$ and deduce that there is a sequence of rational numbers whose limit is $\sqrt{2}$.

6.* Modify the construction in Q5 to produce an *increasing* sequence (b_n) of rational numbers whose limit is $\sqrt{2}$.

7. Let A be a non-empty subset of $\mathbf{R}$ which is bounded above, and define B and C by $B = \{x \in \mathbf{R} : x - 1 \in A\}$ and $C = \{x \in \mathbf{R} : (x + 1)/2 \in A\}$. Prove that $\sup B = 1 + \sup A$, $\sup C = 2 \sup A - 1$.

8. Let A be a non-empty set of *positive* numbers, which is bounded above, and set $B = \{x \in \mathbf{R} : 1/x \in A\}$. Prove that B is bounded below and $\inf B = 1/\sup A$.

9. Suppose that A and B are non-empty bounded subsets of $\mathbf{R}$. Prove that

$$\sup(A \cup B) = \max(\sup A, \sup B),$$

$$\inf\{x + y : x \in A \quad \text{and} \quad y \in B\} = \inf A + \inf B,$$

and

$$\sup\{x - y : x \in A \quad \text{and} \quad y \in B\} = \sup A - \inf B.$$

10. Let A and B be non-empty, bounded sets of *positive* numbers and define C by $C = \{xy : x \in A \text{ and } y \in B\}$. Prove that C is bounded and that $\sup C = \sup A \cdot \sup B$, $\inf C = \inf A \cdot \inf B$.

CHAPTER 7

Continuity

In much the same way as for a sequence, we can consider the behaviour of a function of a real variable x as x tends to some particular number and we should expect to be able to establish the corresponding results without much more effort than that of translating the old terms into the new. Functions of a real variable x, however, may tend to a limit as x tends to any one of infinitely many values in the domain of the function so, as we shall see, the results in this chapter include a second group of properties obtained by collating the properties relating to the individual points of the domain.

Definition   Suppose that, for some $h > 0$, $(c - h, c) \cup (c, c + h)$ belongs to the domain of the function f. We say that $f(x) \rightarrow L$ as $x \rightarrow c$ iff $\forall \varepsilon > 0\ \exists \delta > 0$ s.t. $\forall x \in \mathbf{R}\ 0 < |x - c| < \delta \Rightarrow |f(x) - L| < \varepsilon$.

   Notice a few points about this definition: as with the definition of the limit of a sequence, once we have decided on our 'tolerance' ε, so long as this is positive, there is a positive δ so that if x is near enough c ('near enough' being less than δ from c) then $f(x)$ differs from L by less than ε. We do not consider $x = c$, since we are interested in the behaviour of $f(x)$ as x approaches c; the value of $f(c)$ itself is not relevant—indeed $f(c)$ need not be defined. If $f(x) \rightarrow L$ as $x \rightarrow c$ we write $L = \lim_{x \rightarrow c} f(x)$.

Example   We establish the existence of a limit in much the same way as for sequences. Let $f(x) = x^2$ for all $x \in \mathbf{R}$ so $f : \mathbf{R} \rightarrow \mathbf{R}$. Let $c \in \mathbf{R}$. We prove that $\lim_{x \rightarrow c} f(x) = c^2$.

   Let $\varepsilon > 0$. Then let $\delta = \min(1, \varepsilon/(2|c| + 1))$. Then

$$\forall x \in \mathbf{R} \quad |x - c| < \delta \Rightarrow |f(x) - f(c)| = |x^2 - c^2| = |x - c| \cdot |x + c|$$
$$\leq \delta(|x| + |c|) < \delta(|c| + 1 + |c|) \leq \varepsilon$$

(where we have used the result

$$|x - c| < \delta \Rightarrow |x| \leq |x - c| + |c| < \delta + |c| \leq 1 + |c|).$$

Since $\varepsilon > 0$ was arbitrary, the above is true for all $\varepsilon > 0$, hence

$$\forall \varepsilon > 0 \quad \exists \delta > 0 \quad \text{s.t.} \quad \forall x \in \mathbf{R} \quad |x - c| < \delta \Rightarrow |f(x) - f(c)| < \varepsilon.$$

   It should now be natural to expect the following Lemma to be true, by analogy with what we know of sequences:

71

Lemma 7.1   Let $f(x) \to L$ and $g(x) \to M$ as $x \to c$. Then if λ is a constant, $f(x) + g(x) \to L + M$, $\lambda f(x) \to \lambda L$, $f(x)g(x) \to LM$, $|f(x)| \to |L|$ and, provided $L \neq 0$, $1/f(x) \to 1/L$ as $x \to c$.

Proof: These results are proved by translating the techniques of Theorems 4.1 and 4.2 into the new situation. For definiteness, let us do one and prove the last case. The rough working is as in the corresponding sequence result—we must find a value of δ, say δ_1, such that $|1/f(x)|$ does not exceed some constant if $0 < |x - c| < \delta_1$.

Let $\varepsilon > 0$. Then (since $|f(x)| \to |L|$ which we presume already proved)

$$\exists \delta_1 > 0 \quad \text{s.t.} \quad \forall x \in \mathbf{R} \quad 0 < |x - c| < \delta_1 \Rightarrow \big| \, |f(x)| - |L| \, \big| < \tfrac{1}{2}|L|$$
$$\Rightarrow \tfrac{1}{2}|L| < |f(x)|$$
$$\Rightarrow 1/|f(x)| < 2/|L|. \tag{1}$$

Also (since $f(x) \to L$ and $\tfrac{1}{2}|L|^2 \varepsilon > 0$)

$$\exists \delta_2 > 0 \quad \text{s.t.} \quad \forall x \in \mathbf{R} \quad 0 < |x - c| < \delta_2 \Rightarrow |f(x) - L| < \tfrac{1}{2}L^2 \varepsilon. \tag{2}$$

Then let $\delta = \min(\delta_1, \delta_2)$, so $\delta > 0$ and $|x - c| < \delta \Rightarrow (|x - c| < \delta_1$ and $|x - c| < \delta_2)$. Now, for all $x \in \mathbf{R}$,

$$0 < |x - c| < \delta \Rightarrow \left| \frac{1}{f(x)} - \frac{1}{L} \right| = \frac{|f(x) - L|}{|f(x)| \cdot |L|}$$
$$\leq (2/|L|^2)|f(x) - L| \quad \text{by (1)}$$
$$< \varepsilon \qquad\qquad\qquad \text{by (2)}.$$

Since $\varepsilon > 0$ was arbitrary, the above paragraph is true for all $\varepsilon > 0$ and so $1/f(x) \to 1/L$ as $x \to c$.

The remaining cases are left as an exercise. $\square$

There are two further results about sequences whose analogies we must consider. If we know that $h > 0$ and $\forall x \in (c - h, c) \cup (c, c + h) \ f(x) \leq g(x)$, that $f(x) \to L$ and that $g(x) \to M$ as $x \to c$ then we would expect that $L \leq M$. This result is true, and left as a problem. As with sequences the additional information that $\forall x \in (c - h, c) \cup (c, c + h) \ f(x) < g(x)$ does not, of itself, guarantee $L < M$. The remaining result on sequences is that which claims that if $a_n \to a$ as $n \to \infty$ then (a_n) is bounded. This is not true of functions as the following loose argument suggests: if $a_n \to a$ as $n \to \infty$, then $a - 1 < a_n < a + 1$ for all n 'near ∞', which turns out (when made precise) to be for all but finitely many n. We can then consider the maximum and minimum values of a_n for these finitely many values of n. If $f(x) \to L$ as $x \to c$ then $L - 1 < f(x) < L + 1$ for x 'near c', that is, for $x \in (c - \delta, c) \cup (c, c + \delta)$, but, unlike the case of a sequence, this will usually leave infinitely many values of x for which we have no information about $f(x)$. A function can have 'behaviour' at many points, and it is not hard to find one which tends to a limit as $x \to 1$ but which is nevertheless unbounded; consider $f: (0, \infty) \to \mathbf{R}$ given by $f(x) = 1/x$.

The definition of limit we have given involves the behaviour of $f(x)$ as x approaches c, paying no regard to whether x is greater or less than c. It is not hard to find functions where the behaviour differs on the two sides of c; if we set $f(x) = x/|x|$ (for $x \neq 0$) and $f(0) = 0$ then $f(x) = 1$ for $x > 0$ and $f(x) = -1$ for $x < 0$ and it is obvious that we should consider $x > 0$ and $x < 0$ separately. It is easily checked that $f(x)$ does not tend to a limit as $x \to 0$, since no matter how we choose $\delta > 0$, there are values of x satisfying $|x| < \delta$ for which $f(x) = +1$ and others for which $f(x) = -1$. We therefore define 'one-sided' limits.

Definition   Let $f:(c, c + h) \to \mathbf{R}$ for some $h > 0$. Then we say $f(x)$ tends to L as x tends to c from the right, or $f(x) \to L$ as $x \to c+$, iff

$$\forall \varepsilon > 0 \quad \exists \delta > 0 \quad \text{s.t.} \quad \forall x \in (c, c + \delta) \quad |f(x) - L| < \varepsilon.$$

When this holds we write $L = \lim_{x \to c+} f(x)$. If $f:(c - h, c) \to \mathbf{R}$ for some $h > 0$ then we define $\lim_{x \to c-} f(x)$, the limit as x tends to c from the left, in the analogous way.

Example   If $f(x) = x/|x|$ for $x \in \mathbf{R} \backslash \{0\}$, then $\lim_{x \to 0+} f(x) = 1$ and $\lim_{x \to 0-} f(x) = -1$.

Solution: Let $\varepsilon > 0$. Then $\exists \delta > 0$ (e.g. $\delta = 1$) such that $\forall x \in (0, \delta) \, |f(x) - 1| = 0 < \varepsilon$. Since $\varepsilon > 0$ was arbitrary we have proved that $f(x) \to 1$ as $x \to 0+$. The proof for the left-hand limit is similar.

Example   Let $f:(c - h, c) \cup (c, c + h) \to \mathbf{R}$, for some $h > 0$. Then $\lim_{x \to c} f(x)$ exists if and only if $\lim_{x \to c-} f(x)$ and $\lim_{x \to c+} f(x)$ both exist and are equal.

Solution: Suppose $\lim_{x \to c} f(x)$ exists and that its value is L. Then

$$\forall \varepsilon > 0 \quad \exists \delta > 0 \quad \text{s.t.} \quad \forall x \in \mathbf{R} \quad 0 < |x - c| < \delta \Rightarrow |f(x) - L| < \varepsilon.$$

Since $x \in (c, c + \delta) \Rightarrow x \in \mathbf{R}$ and $0 < |x - c| < \delta$ it is easy to see that $\lim_{x \to c+} f(x) = L$ and similarly $\lim_{x \to c-} f(x) = L$.

   Conversely, suppose that $\lim_{x \to c-} f(x)$ and $\lim_{x \to c+} f(x)$ both exist and have the common value L. Let $\varepsilon > 0$. Then

$$\exists \delta_1 > 0 \quad \text{s.t.} \quad \forall x \in (c - \delta_1, c) \quad |f(x) - L| < \varepsilon$$

and

$$\exists \delta_2 > 0 \quad \text{s.t.} \quad \forall x \in (c, c + \delta_2) \quad |f(x) - L| < \varepsilon.$$

Therefore, letting $\delta = \min(\delta_1, \delta_2)$, $\forall x \in \mathbf{R}$ $0 < |x - c| < \delta$ implies that $x \in (c - \delta_1, c)$ or $x \in (c, c + \delta_2)$, and in either case $|f(x) - L| < \varepsilon$. Thus $\exists \delta > 0$ s.t. $\forall x \in \mathbf{R}$ $0 < |x - c| < \delta \Rightarrow |f(x) - L| < \varepsilon$.
   Since $\varepsilon > 0$ was arbitrary we have proved that $\lim_{x \to c} f(x) = L$.

   An easy modification to the proof shows that Lemma 7.1 remains true if left-hand limits are substituted throughout for 'two-sided' limits or if right-hand limits are substituted throughout. The one result we gain is the

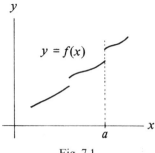

Fig. 7.1

analogue of the result guaranteeing the existence of a limit of an increasing sequence which is bounded above.

Example   Suppose $f:\mathbf{R} \to \mathbf{R}$ is increasing, that is, $x \le y \Rightarrow f(x) \le f(y)$. Then for all $a \in \mathbf{R}$, $\lim_{x \to a-} f(x)$ exists and equals $\sup \{f(x):x < a\}$. (See Fig. 7.1.)

Solution: Since f is increasing, $f(a)$ is an upper bound for $\{f(x):x < a\}$ so this set has a supremum; call the supremum α.

Let $\varepsilon > 0$. Then since $\alpha - \varepsilon < \alpha$, from the properties of supremum, $\exists x_0 < a$ for which $f(x_0) > \alpha - \varepsilon$. Let $\delta = a - x_0$, so $\delta > 0$ and $x_0 = a - \delta$, and, using the increasing property of f, $\forall x \in (a - \delta, a)$ $\alpha - \varepsilon < f(x_0) \le f(x) \le \alpha$, whence $|f(x) - \alpha| < \varepsilon$. Since $\varepsilon > 0$ was arbitrary this shows us that $\forall \varepsilon > 0$ $\exists \delta > 0$ s.t. $\forall x \in (a - \delta, a)$ $|f(x) - \alpha| < \varepsilon$, that is, that $\lim_{x \to a-} f(x) = \alpha$.

The most important connection in which we use limits is where $\lim_{x \to a} f(x) = f(a)$. A function with this property at all points a is one whose graph can be drawn without 'jumps' and 'without taking the pencil off the paper'. The property is so common we give it a name.

Definition   Let f be a function whose domain includes $(c - h, c + h)$ for some $h > 0$. Then we say f is **continuous at** c if and only if $\lim_{x \to c} f(x) = f(c)$. A function is said to be **continuous** iff it is continuous at each point of its domain; this is only a matter of applying the definition just given if the domain is a set of one of the forms $\mathbf{R}$, $(-\infty, b)$, (a, ∞) or (a, b), but if $f:[a, b] \to \mathbf{R}$ we have to make a special case at the two endpoints (where, for all $h > 0$, f is not, for example, defined on $(a - h, a + h)$). A function $f:[a, b] \to \mathbf{R}$ is said to be continuous iff it is continuous at each point of (a, b) and $\lim_{x \to a+} f(x) = f(a)$ and $\lim_{x \to b-} f(x) = f(b)$.

We shall use continuity so much that it is worth while to restate it in terms of ε and δ: f is continuous at c if and only if

$$\forall \varepsilon > 0 \quad \exists \delta > 0 \quad \text{s.t.} \quad \forall x \in (c - \delta, c + \delta) \quad |f(x) - f(c)| < \varepsilon.$$

Notice that we do not need to exclude the case $x = c$ since the limit in this case is $f(c)$ and $|f(c) - f(c)| = 0 < \varepsilon$. The continuity conditions at a and b in

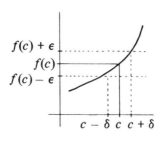

Fig. 7.2

the case of $f:[a,b]\to\mathbf{R}$ are called continuity on the right and left respectively. (See Fig. 7.2.)

Lemma 7.1 may be applied to show that if f and g are continuous functions and λ is a constant, then $\lambda f, f + g, fg$ and $|f|$ are all continuous; by fg we mean the function whose value at x is the product $f(x)g(x)$ of the values of f and g. The function $1/g$ is continuous at all points c for which $g(c)\neq 0$. There is, however, one operation on functions which has not arisen before, that of composition.

Definition   Let $f:B\to C$ and $g:A\to B$ be two functions. The **composition** of f and g, $f\circ g$, is the function $f\circ g:A\to C$ defined by $(f\circ g)(x) = f(g(x))$.

Lemma 7.2   Let $f:B\to C$ and $g:A\to B$ be functions, where A, B and C are subsets of $\mathbf{R}$. If g is continuous at a and f is continuous at $b = g(a)$, then $f\circ g$ is continuous at a. If f and g are continuous, so is $f\circ g$.

Proof: Suppose that $b = g(a)$, f is continuous at b and g is continuous at a. Let $\varepsilon > 0$.

Since f is continuous at b,

$$\exists\delta_1 > 0\quad\text{s.t.}\quad |y - b| < \delta_1 \Rightarrow |f(y) - f(b)| < \varepsilon.$$

For this value of δ_1, since $\delta_1 > 0$ and g is continuous at a,

$$\exists\delta_2 > 0\quad\text{s.t.}\quad |x - a| < \delta_2 \Rightarrow |g(x) - g(a)| < \delta_1.\quad\text{Then}$$
$$|x - a| < \delta_2 \Rightarrow |g(x) - b| < \delta_1 \Rightarrow |f(g(x)) - f(b)|$$
$$= |(f\circ g)(x) - (f\circ g)(a)| < \varepsilon.$$

Since $\varepsilon > 0$ was arbitrary, we have shown $f\circ g$ is continuous at a.   □

The results about continuous functions allow us to deduce immediately that many standard functions are continuous. We know that a constant function and the function $g:\mathbf{R}\to\mathbf{R}$ defined by $g(x) = x$ are continuous, so since any polynomial can be obtained from these by a finite number of operations of adding, multiplying by a constant or multiplying two functions, we see that polynomials are continuous. Any function which is the quotient of two polynomials is continuous at all points where the denominator is non-zero.

The importance of continuous functions is rather more substantial than we might at this point expect, in that we can deduce 'global' properties of these functions which derive from the fact that they are continuous at all points of their domain. These results have an additional level of sophistication to the proof in which we bring the underlying properties of the real number system into play.

Theorem 7.3 (The Intermediate Value Theorem)   Suppose that $f:[a,b] \to \mathbf{R}$ is continuous and that γ lies between $f(a)$ and $f(b)$, in the sense that either $f(a) \le \gamma \le f(b)$ or $f(a) \ge \gamma \ge f(b)$. Then there is a number $\xi \in [a,b]$ for which $f(\xi) = \gamma$.

Remarks: This result is 'obvious' in the sense that it is easily believed, so easily that one can scarcely conceive that it could be false. The fact that one cannot imagine how it could be wrong is, of course, not a proof.

Proof: We can reduce this to a more particular case by noticing that if f is continuous, so is g, where $g(x) = f(x) - \gamma$, and then 0 lies between $g(a)$ and $g(b)$. By this device we need only prove the case when $\gamma = 0$. By considering $-g$ in place of g we could deduce the result for the case $g(a) \ge 0 \ge g(b)$ from that where $g(a) \le 0 \le g(b)$. Finally, if either $g(a)$ or $g(b)$ is zero, there is no trouble in finding a ξ for which $g(\xi) = 0$. We may therefore deduce all we wish if we can show that $f(a) < 0 < f(b)$ implies the existence of $\xi \in [a,b]$ for which $f(\xi) = 0$.

The way forward is to identify a candidate for ξ; from Fig. 7.3 we choose $\xi = \sup S$, where $S = \{x \in [a,b] : f(x) < 0\}$. The set S is non-empty (it contains a) and bounded above (by b) so ξ exists and $a \le \xi \le b$. We have to show $f(\xi) = 0$.

Suppose $f(\xi) > 0$. Then, since f is continuous at ξ, $\exists \delta > 0$ s.t. $\forall x \in (\xi - \delta, \xi + \delta)$ $|f(x) - f(\xi)| < f(\xi)$. (Setting $\varepsilon = f(\xi)$.) Since $|f(x) - f(\xi)| < f(\xi) \Rightarrow f(\xi) - f(\xi) < f(x) < f(\xi) + f(\xi)$ we deduce that $\forall x \in (\xi - \delta, \xi + \delta)$ $f(x) > 0$. Since $\xi - \delta < \sup S$, $\exists y > \xi - \delta$ s.t. $y \in S$, that is, such that $f(y) < 0$. But this is a contradiction since $y \in (\xi - \delta, \xi + \delta) \Rightarrow f(y) > 0$. From this contradiction it follows that $f(\xi) \not> 0$.

Now suppose $f(\xi) < 0$. Then by the continuity of f at ξ, and since $-f(\xi) > 0$, $\exists \delta > 0$ s.t. $\forall x \in (\xi - \delta, \xi + \delta)$ $|f(x) - f(\xi)| < -f(\xi)$ whence

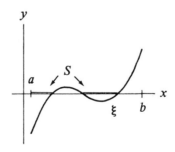

Fig. 7.3

$\forall x \in (\xi - \delta, \xi + \delta) \; f(x) < f(\xi) - f(\xi) = 0$. Therefore $f(\xi + \frac{1}{2}\delta) < 0$ so $\xi + \frac{1}{2}\delta \in S$ and $\xi + \frac{1}{2}\delta > \sup S$, a contradiction. This proves that $f(\xi) \not< 0$.

We are left only with the possibility $f(\xi) = 0$, which must be true. □

Example This result allows us to show with virtually no effort that nth roots exist. Let a be positive and $n \in \mathbf{N}$. The function $f : [0, \infty) \to \mathbf{R}$ defined by $f(x) = x^n$ is continuous, and $f(0) < a$. If $a \le 1$ then $f(0) \le a \le f(1)$, while if $a > 1$ then $a^n \ge a$ so $f(0) \le a \le f(a)$. Applying the Intermediate Value Theorem (IVT) we see that there is a ξ for which $\xi^n = a$.

The IVT may also be deployed to give information on the existence of solutions of more complicated equations. One could show the existence of a solution to the equation $x^2 + 7x = x^3 + 1$ by showing that the (continuous) function f given by $f(x) = x^3 - x^2 - 7x + 1$ attains a positive value at $x = 0$ and a negative one at $x = -3$, and is thus zero at some intermediate point. There are obviously some practical matters to be attended to before this can be made greatly useful.

Definition Let $f : A \to \mathbf{R}$. f is said to be **bounded** if the set $\{f(x) : x \in A\}$ is a bounded set. For $S \subset A$, we say f is **bounded on** S if $\{f(x) : x \in S\}$ is bounded. The analogous definitions of bounded above and below are also made.

Thus $f : \mathbf{R} \to \mathbf{R}$ defined by $f(x) = 1/(1 + x^2)$ is bounded, but $g : (0, 1) \to \mathbf{R}$ given by $g(x) = 1/x$ is not.

Experience suggests that if $f : [a, b] \to \mathbf{R}$ is continuous then f is bounded. This we shall prove shortly, but as a preliminary, let us isolate the difficulties. For each $c \in [a, b]$, since f is continuous at c, $\exists \delta_c > 0$ such that $\forall x \in (c - \delta_c, c + \delta_c) \cap [a, b] \; f(c) - 1 < f(x) < f(c) + 1$. (We adopt this formulation to avoid exceptions at the end points.) Thus the function obtained by restricting f to the domain $(c - \delta_c, c + \delta_c) \cap [a, b]$ is bounded. One natural approach would be to choose a finite number of points $c_1, \ldots, c_n$ in such a way that every point of $[a, b]$ lies in one of the intervals $(c - \delta_c, c + \delta_c)$. It is here that care is needed; δ_c may depend on c. Once we know the point c and that f is continuous at c, continuity guarantees the existence of $\delta_c > 0$, but the suitable values of δ may not be the same at another point c'. This prevents the simple approach that would be possible if δ_c were a constant, when it would be easy to find a finite number of intervals of fixed length covering $[a, b]$. If we require infinitely many subintervals, the upper bounds for the various subintervals need not automatically form a bounded set.

Theorem 7.4 Let $f : [a, b] \to \mathbf{R}$ be continuous. Then f is bounded.

Proof: We introduce the set $S = \{x \in [a, b] : f \text{ is bounded on } [a, x]\}$. Our problem is to show that $b \in S$. S is bounded above (by b) and non-empty (it contains a) so S has a supremum. Let $\xi = \sup S$.

Suppose $\xi < b$. Then since f is continuous at ξ, $\exists \delta > 0$ s.t. $\forall y \in (\xi - \delta, \xi + \delta) \cap [a, b], \; f(\xi) - 1 < f(y) < f(\xi) + 1$. By reducing δ if necessary we may presume $\xi + \delta \le b$. Moreover, since $\xi - \delta < \sup S$, $\exists x \in S$ with $\xi - \delta < x \le \xi$. Then f is bounded on $[a, x]$ so, for some constants m, M, we have $\forall y\in[a,x]\ m\leq f(y)\leq M$. Hence $\forall y\in[a,\xi+\frac{1}{2}\delta]\ \min(m,f(\xi)-1)\leq f(y)\leq \max(M,f(\xi)+1)$ (since $a\leq y\leq \xi+\frac{1}{2}\delta\Rightarrow a\leq y\leq x$ or $\xi-\delta<y<\xi+\delta$). Thus f is bounded on $[a,\xi+\frac{1}{2}\delta]$ and $\xi+\frac{1}{2}\delta\in S$. But $\xi+\frac{1}{2}\delta>\sup S$. This contradiction shows $\xi\nless b$.

Therefore $\xi=b$ (since b is an upper bound for $S,\xi\leq b$). By the continuity of f on the left at b, $\exists\delta>0$ s.t. $\forall y\in(b-\delta,b]\ f(b)-1<f(y)<f(b)+1$. Moreover since $b-\delta<\sup S,\exists x\in S$ with $b-\delta<x$. Thus $\exists m,M$ s.t. $\forall y\in[a,x]$ $m\leq f(y)\leq M$ whence $\forall y\in[a,b]\ \min(m,f(b)-1)\leq f(y)\leq \max(M,f(b)+1)$, that is, f is bounded on $[a,b]$. $\square$

Remarks: It is worth considering this proof carefully in the light of examples. If we allow f to be defined on an open interval (a,b) the result is false (e.g. $f(x)=1/x$ on $(0,1)$), or on an infinite interval (e.g. $f(x)=x$ on $[0,\infty)$).

Theorem 7.5 Let $f:[a,b]\to\mathbf{R}$ be continuous. Then f attains its supremum and infimum, that is, there are numbers ξ_1 and ξ_2 in $[a,b]$ for which $f(\xi_1)=\sup\{f(x):a\leq x\leq b\}$ and $f(\xi_2)=\inf\{f(x):a\leq x\leq b\}$.

Proof: Let $M=\sup\{f(x):a\leq x\leq b\}$. (By Theorem 7.4 we know M exists.) Suppose that $\forall x\in[a,b]\ f(x)\neq M$ so that $\forall x\in[a,b]\ M-f(x)>0$. Then set $g(x)=1/(M-f(x))$ so $g:[a,b]\to\mathbf{R}$ and g is continuous. By Theorem 7.4, then, g is bounded, so $\exists M_1$ s.t. $\forall x\in[a,b]\ g(x)\leq M_1$. Then $\forall x\in[a,b]$ $M-f(x)\geq 1/M_1$ and $f(x)\leq M-1/M_1$. But this shows that $M-1/M_1$ is an upper bound for f which is smaller than the supremum of f, a contradiction. Thus our assumption that $\forall x\in[a,b]\ f(x)\neq M$ was wrong, so $\exists\xi_1\in[a,b]$ s.t. $f(\xi_1)=M$. The other case is now clear. $\square$

Example One useful conclusion from this is that if $f:[a,b]\to\mathbf{R}$ is continuous and all its values are positive, then its infimum is positive. Before showing this we ought to say why we bother with it: a set of positive numbers can have 0 as its infimum, for example $(0,1)$ does, so it is conceivable that f could have lots of local minima in the interval $[a,b]$, all positive, but including some arbitrarily close to 0.

Solution: By the theorem $f:[a,b]\to\mathbf{R}$ attains its infimum, so $\exists\xi\in[a,b]$ for which $f(\xi)=\inf\{f(x):a\leq x\leq b\}$. Since all the values of f are positive, $f(\xi)>0$.

Notice that our results may be useful even with functions continuous on sets other than closed intervals, provided we are careful. For example, if $f:(0,\infty)\to\mathbf{R}$ is continuous, and all its values are positive then $\{f(x):a\leq x\leq b\}$ has a positive infimum provided $0<a<b<\infty$, since the restriction of f to $[a,b]$ is continuous.

In everyday life it is natural to presume that in whatever process we are concerned with, a small change in the input will produce a small change in the output; if instructions call for ingredients to be mixed in the proportions 2:1 we do not interpret this as demanding that there be exactly twice as many molecules of the one as the other; we approximate this. Again, common experience tells us that with some processes we need to be more accurate in our approximation than with others. Continuity is essentially the same idea: if x is close to a then $f(x)$ will be close to $f(a)$; the idea of how close is necessary to attain a certain accuracy being measured by ε and δ. Having said that continuity is natural, we need to pursue two distinct approaches: to try to check that continuity is present where we would expect it and to be aware of where it is absent.

We have shown that the sum, product and composition of two continuous functions are continuous, and that the quotient of two continuous functions is continuous except where the denominator is zero. These results are all of the general form that performing the appropriate operations on two continuous functions yields a result which is continuous. There are, however, other ways in which we construct functions out of other functions; for example, the square root function is obtained from the simpler function which maps x into x^2 by 'undoing' the action of the latter, or, in more formal terms, the square root function is the 'inverse' of the square function. Many other standard functions of mathematics are inverses of simpler functions, so we must relate the operation of inverting to continuity. We start by defining our terms.

Definition Suppose that A and B are two sets and $f:A\rightarrow B$. Then we say that $g:B\rightarrow A$ is the **inverse function** to f if $\forall x\in A$ $g(f(x))=x$ and $\forall y\in B$ $f(g(y))=y$. If such an inverse function g exists, we denote it by f^{-1} (not to be confused with $1/f$).

$f:A\rightarrow B$ is said to be **injective** if $\forall x_1, x_2\in A$ $x_1\neq x_2\Rightarrow f(x_1)\neq f(x_2)$ (or, equivalently, $f(x_1)=f(x_2)\Rightarrow x_1=x_2$).

Examples Let $f_1:[0,\infty)\rightarrow[0,\infty)$ be given by $f_1(x)=x^2$ and $g_1:[0,\infty)\rightarrow[0,\infty)$ be given by $g_1(x)=\sqrt{x}$. Then $g_1=f_1^{-1}$. Notice also that f_1 is injective, while $f_2:\mathbf{R}\rightarrow[0,\infty)$ given by $f_2(x)=x^2$ is not injective ($f_2(1)=f_2(-1)$). Mark this example: the domain of the function is as important as the rule for transforming x to $f(x)$.

We are interested in conditions which guarantee the existence of an inverse function. We first notice that if f has an inverse (g, say) then f is injective, for, in the notation of the definition, $f(x_1)=f(x_2)\Rightarrow g(f(x_1))=g(f(x_2))\Rightarrow x_1=x_2$. The second point is that if $f:A\rightarrow B$ is to have an inverse then every point of B must be in the image of f (since $y=f(g(y))$ in the above notation). In practice, we often pay less attention to the codomain so we may be prepared to alter it. Let $f(A)$ denote the set $\{f(x):x\in A\}$ when $f:A\rightarrow B$. Then if f is injective, we see that for all $y\in f(A)$ there is an $x\in A$ satisfying $f(x)=y$ and, since f is injective, there is only one such x. Define $g(y)$ to be x. Then $g:f(A)\rightarrow A$ and it is easily checked that g is the inverse of f. We have proved:

Lemma 7.6 Let $f:A\rightarrow B$. Then the function $f:A\rightarrow f(A)$ has an inverse function $g:f(A)\rightarrow A$ if and only if f is injective. Notice that the domain of g need not be all of B.

From the point of view of analysis, we shall be interested in showing that the inverse of a continuous function, if it exists, is continuous. The first step is to notice that the only injective continuous functions are the obvious ones.

Definitions Let $A \subset \mathbf{R}$ and $f:A \to \mathbf{R}$. f is said to be **increasing** if $x < y \Rightarrow f(x) \le f(y)$ and **strictly increasing** if $x < y \Rightarrow f(x) < f(y)$. We say f is **decreasing (strictly decreasing)** if $x < y \Rightarrow f(x) \ge f(y)$ ($f(x) > f(y)$).

Lemma 7.7 Let $f:[a,b] \to \mathbf{R}$ be continuous and injective. Then f is either strictly increasing or strictly decreasing.

Proof: Since f is injective, $f(a) \ne f(b)$ (neglecting the trivial case $a = b$). Thus either $f(a) < f(b)$ or $f(a) > f(b)$. We shall tackle the former case; the other is nearly identical. Suppose, therefore, that $f(a) < f(b)$.

Let $a \le x < y \le b$. Then $f(a) \le f(x) < f(b)$. For, if not, either $f(x) < f(a)$ or $f(x) \ge f(b)$. In the first case, $f(x) < f(a) < f(b)$, so $f(a)$ is intermediate between $f(x)$ and $f(b)$, and, applying the IVT to f on the interval $[x,b]$, we see that $\exists \xi \in (x,b)$ s.t. $f(\xi) = f(a)$. But then $\xi \ne a$, which contradicts the injectivity of f. (Note $\xi \ne x$ or b since $f(\xi)$ does not equal $f(x)$ or $f(b)$.) Thus $f(x) \not< f(a)$. Similarly if $f(x) \ge f(b)$, we have $f(a) < f(b) \le f(x)$ so, applying the IVT to $[a,x]$ we see that $\exists \xi \in [a,x]$ with $f(\xi) = f(b)$. Since $\xi \le x < b$, this contradicts the injectivity of f.

Moreover, $f(x) < f(y)$, for f is continuous and injective on $[x,b]$, so if $f(y) \le f(x)$ we have $f(y) \le f(x) < f(b)$ (since we know $f(x) < f(b)$) whence applying the IVT to $[y,b]$ shows $\exists \xi \in [y,b]$ s.t. $f(\xi) = f(x)$. Since $\xi \ne x$ this is impossible. Thus $f(x) < f(y)$.

Since x and y were typical points of $[a,b]$ with $x < y$ we have proved that $a \le x < y \le b \Rightarrow f(x) < f(y)$, so f is strictly increasing on $[a,b]$. $\square$

Theorem 7.8 Let $f:[a,b] \to \mathbf{R}$ be continuous and injective. Then $f^{-1}: f([a,b]) \to [a,b]$ is continuous.

Proof: Since f is continuous and injective it is either strictly increasing or strictly decreasing. We shall prove the former case. Then $x \in [a,b] \Rightarrow f(x) \in [f(a), f(b)]$. In fact, the IVT allows us to show that every number between $f(a)$ and $f(b)$ belongs to $f([a,b])$.

Let $f(a) \le y_1 < y_2 \le f(b)$, and let $x_i = f^{-1}(y_i)$ ($i = 1, 2$). Since f is strictly increasing, $x_1 < x_2$ (since $x_1 \ge x_2 \Rightarrow f(x_1) \ge f(x_2) \Rightarrow y_1 \ge y_2$). Thus f^{-1} is strictly increasing.

Let $y_0 \in (f(a), f(b))$ and $f(x_0) = y_0$. Let $\varepsilon > 0$. We shall assume that ε is sufficiently small that $x_0 \pm \varepsilon \in [a,b]$; if this is not so, we choose a smaller ε for which it is true. Then, since f is strictly increasing, $f(x_0 - \varepsilon) < y_0 < f(x_0 + \varepsilon)$. Let δ be the minimum of $y_0 - f(x_0 - \varepsilon)$ and $f(x_0 + \varepsilon) - y_0$. Then, since f^{-1} is strictly increasing,

$$y \in (y_0 - \delta, y_0 + \delta) \Rightarrow f(x_0 - \varepsilon) < y < f(x_0 + \varepsilon)$$
$$\Rightarrow x_0 - \varepsilon < f^{-1}(y) < x_0 + \varepsilon$$

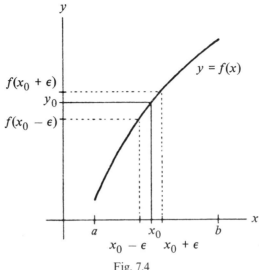

Fig. 7.4

so $y \in (y_0 - \delta, y_0 + \delta) \Rightarrow |f^{-1}(y) - f^{-1}(y_0)| < \varepsilon$. Since $\varepsilon > 0$ was arbitrary we have shown f^{-1} is continuous at y_0, and since y_0 was arbitrary, f^{-1} is continuous on $(f(a), f(b))$. Minor changes in the proof above show the continuity at the endpoints. □

Applying this result to the function $f:[0, b] \rightarrow [0, b^n]$ given by $f(x) = x^n$ (where n is a given natural number) shows that the nth root function is continuous on $[0, b^n]$. To show the root function is continuous on all of $[0, \infty)$ we need only notice that if $a \geq 0$ then it is possible to choose b so that $b^n > a$, hence $x \mapsto x^{1/n}$ is continuous on $[0, b^n]$ and, in particular, at a.

Partly as a warning, partly to see exactly how powerful the results we have proved are, we shall give some examples where continuity is absent. These functions are pathological to our accustomed way of thinking and are not the sort of function we would expect to encounter, though if we produce functions by ingenious processes other than actually naming their values explicitly we need to ensure that such oddities are not present. It is instructive to look at proofs of ordinary results to see how such peculiarities are excluded.

Examples   1. *A function* $f_1 : \mathbf{R} \rightarrow \mathbf{R}$ *which is nowhere continuous.*
Define f_1 by

$$f_1(x) = \begin{cases} 1 & \text{if } x \text{ is rational,} \\ 0 & \text{if } x \text{ is irrational.} \end{cases}$$

To see that f_1 is discontinuous everywhere, let $c \in \mathbf{R}$ and $\varepsilon = 1$. Then for all $\delta > 0$, $(c - \delta, c + \delta)$ contains both rational and irrational numbers, and in particular an x of the opposite type to c; for this x, $|f_1(x) - f_1(c)| = 1 \geq \varepsilon$. Thus $\exists \varepsilon > 0$ s.t. $\forall \delta > 0 \ \exists x \in (c - \delta, c + \delta)$ s.t. $|f_1(x) - f_1(c)| \geq \varepsilon$.

2. *A function* $f_2 : \mathbf{R} \rightarrow \mathbf{R}$ *which is continuous at infinitely many points but between*

every two points of continuity there is a point of discontinuity and vice versa.
Let

$$f_2(x) = \begin{cases} 0 & \text{if} \quad x \text{ is irrational,} \\ 1/q & \text{if} \quad x = p/q \text{ where } p \text{ and } q \text{ are integers with} \\ & \qquad \text{no common factor and } q > 0. \end{cases}$$

Let c be rational, where $f_2(c) = 1/q$. Let $\varepsilon = 1/q(>0)$. Then $\forall \delta > 0$ $\exists x \in (c - \delta, c + \delta)$ (e.g. any irrational x) s.t. $f_2(x) = 0$, hence $|f_2(x) - f_2(c)| = 1/q \geq \varepsilon$. This shows f_2 is discontinuous at c and hence at all rational numbers.

Now suppose c is irrational; we prove f_2 is continuous at c. Let $\varepsilon > 0$. Then if $q \in \mathbf{N}$, $1/q \geq \varepsilon \Leftrightarrow q \leq 1/\varepsilon$. The interval $(c - 1, c + 1)$ contains finitely many rational numbers with a denominator at most $1/\varepsilon$, and since none of these numbers is c, because c is irrational, the nearest of these is a positive distance δ from c. Thus if $|x - c| < \delta$, either x is irrational or x is rational and of the form p/q (where p and q have no common factor) with $q > 1/\varepsilon$, whence $|f_2(x) - f_2(c)| < \varepsilon$. Since ε was arbitrary, f_2 is continuous at c.

3. *A function $f_3: (0, 1] \to \mathbf{R}$ which is continuous and bounded but which does not tend to a limit as $x \to 0+$.*

In this example we shall jump ahead and presume the properties of the sine function. Since none of our theory will depend on this, there is no danger of circular arguments resulting.
Let

$$f_3(x) = \sin(1/x) \quad (x \in (0, 1]).$$

Then (presuming $\sin y$ to be continuous in y) we see that f_3 is the composition of two continuous functions. Also $\forall x \in (0, 1]$ $|f_3(x)| \leq 1$ so f_3 is bounded. Suppose that $f_3(x) \to L$ as $x \to 0+$. Let $\varepsilon > 0$. Then $\exists \delta > 0$ s.t. $\forall x \in (0, \delta)$ $|f_3(x) - L| < \varepsilon$. Now choose $n \in \mathbf{N}$ s.t. $1/n < 2\delta \pi$, which is certainly possible, and choose $x = 1/(2n\pi + \pi/2)$. Then $x \in (0, \delta)$ and $f_3(x) = \sin(2n\pi + \pi/2) = 1$, so $|1 - L| < \varepsilon$. Since $\varepsilon > 0$ was arbitrary we see that $\forall \varepsilon > 0$, $|1 - L| < \varepsilon$, so $|1 - L| \leq 0$ and hence $L = 1$. Again let $\varepsilon > 0$ so $\exists \delta > 0$ s.t. $0 < x < \delta \Rightarrow |f_3(x) - L| < \varepsilon$. Choose $n \in \mathbf{N}$ s.t. $1/n < 2\pi\delta$ and let $x = 1/(2n\pi + 3\pi/2)$, so $0 < x < \delta$, $f_3(x) = -1$ and hence $|-1 - L| < \varepsilon$. Since this is true for all $\varepsilon > 0$ we now deduce $L = -1$. This is a contradiction, so our assumption that $f_3(x)$ tends to a limit as $x \to 0+$ is wrong.

Problems

1. Prove directly from the definition that $\lim\limits_{x \to a} x^3 = a^3$ and
$$\lim\limits_{x \to 0} (1 + x^2)/(2 + x) = 1/2.$$

2. Show that if $f(x) \to L$ and $g(x) \to M$ as $x \to a$, then $f(x)g(x) \to LM$ as $x \to a$, and prove the remaining parts of Lemma 7.1.

3. Suppose that $\forall x \in (-1, 1)$, $1 - |x| \leq f(x) \leq 1 + x^2$. Prove that $f(x) \to 1$ as $x \to 0$.

4. Let f and h be two continuous functions on $(a-1, a+1)$ and suppose that $\forall x \in (a-1, a+1)$, $f(x) \leq g(x) \leq h(x)$. Show that if $f(a) = h(a)$ then g is continuous at a. (Draw a diagram!)

5. Suppose that K is a constant and that $\forall x \in \mathbf{R}$ $|f(x)| \leq K|x|$. Prove that f is continuous at 0. (First find $f(0)$.)

6. Let $f: \mathbf{R} \to \mathbf{R}$ be continuous at a and $f(a) > 0$. Show that there is a positive δ such that $|x-a| < \delta \Rightarrow f(x) > 0$. (Draw a diagram to see how to choose ε.)

7. We say $f(x) \to L$ as $x \to \infty$ iff $\forall \varepsilon > 0 \exists X$ s.t. $\forall x \in \mathbf{R}$ $x \geq X \Rightarrow |f(x) - L| < \varepsilon$. Prove that $1/x \to 0$ as $x \to \infty$, and, working directly from the definition, show that if $f(x) = 1 + a_2/x + a_1/x^2 + a_0/x^3$ (a_0, a_1, a_2 being constants) then $f(x) \to 1$ as $x \to \infty$. Deduce that $\exists X_1$ s.t. $\forall x \geq X_1$ $f(x) > 0$ and hence $\exists R_1$ for which $\forall x \geq R_1$ $x^3 + a_2 x^2 + a_1 x + a_0 > 0$. By considering $g(x) = 1 - a_2/x + a_1/x^2 - a_0/x^3$ as $x \to \infty$, show that $\exists R_2$ s.t. $\forall x \geq R_2$ $x^3 - a_2 x^2 + a_1 x - a_0 > 0$ to deduce that $\forall y \leq -R_2$ $y^3 + a_2 y^2 + a_1 y + a_0 < 0$. Finally, use the IVT to prove that the equation $x^3 + a_2 x^2 + a_1 x + a_0 = 0$ has at least one real solution.

8. Let $f: [0, 1] \to [0, 1]$ be continuous. By considering the function h given by $h(x) = f(x) - x$, or otherwise, deduce that there is a $\xi \in [0, 1]$ for which $f(\xi) = \xi$.

9. Suppose that $g: \mathbf{R} \to \mathbf{R}$ is continuous and that $\forall x \in \mathbf{Q}$ $g(x) = 0$. Deduce that g is identically zero.

10. Let $f: [a, b] \to \mathbf{R}$ be continuous. Prove that $\{f(x): a \leq x \leq b\}$ is a closed, bounded interval.

11.* Suppose that $f: [0, \infty) \to \mathbf{R}$ is continuous and that $f(x) \to L$ as $x \to \infty$ (definition in Q7). Show that f is bounded. (Hint: Use the limit to establish bounds for $\{f(x): x > X\}$, for a suitable X, then deal with $\{f(x): 0 \leq x \leq X\}$.)

12.* Let $f: \mathbf{R} \to \mathbf{R}$ be continuous and bounded, and define g by $g(x) = \sup\{f(x): y \leq x\}$. Prove that g is continuous.

13.* Let $h: \mathbf{R} \to \mathbf{R}$ be continuous and neither bounded above nor below. Show that $\{h(x): x \in \mathbf{R}\} = \mathbf{R}$.

CHAPTER 8

Differentiation

"Just as from the known *Algorithm*, as I call it, of this calculus, which I call *differential*, all other differential equations can be solved by a common method...".

G. W. Leibniz (1684).

Curiously, although calculus is about differentiation and it was the need to put calculus on a rigorous foundation that gave rise to analysis, the substantial work has already been done. This will allow us to make use of our previous efforts to dispose rather quickly of the standard results, though our viewpoint will make it more natural to state some results which, though useful, would seem odd had we approached them from a background purely of calculus.

The first thing to be done is to define the derivative:

Definition Let f be a real-valued function whose domain includes $(c-h, c+h)$ for some $h > 0$. Then f is said to be **differentiable at** c if $\lim_{x \to c} (f(x) - f(c))/(x - c)$ exists, and when it does we denote its value by $f'(c)$, and call $f'(c)$ the **derivative** of f at c. A function $f : S \to \mathbf{R}$ is said to be **differentiable** if it is differentiable at every point of S.

This definition does not allow us to consider the differentiability of f at a if $f : [a, b] \to \mathbf{R}$; if the need arises we demand that the obvious one-sided limit exists.

If $f : S \to \mathbf{R}$ is differentiable, then for all $c \in S$ we have defined $f'(c)$, so that $f' : S \to \mathbf{R}$ is a new function also called the derivative of f. It can be important

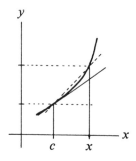

Fig. 8.1. $f'(c)$, being the limit of the gradient of the straight line from $(c, f(c))$ to $(x, f(x))$ as $x \to c$, is the gradient of the tangent to the curve $y = f(x)$ at c.

to distinguish between the function f' and its value at the point c, $f'(c)$, and this notation (Newton's) does this conveniently. The alternative notation (Leibniz's) of df/dx is rather less well suited to making this distinction, though it has a number of other advantages, particularly that it suggests many correct results. If the function f' is differentiable at c we say f is twice differentiable there and denote the derivative of f' at c by $f''(c)$; if f'' is differentiable we denote its derivative by f''' or $f^{(3)}$, and so on.

Theorem 8.1   If f is differentiable at c it is continuous at c.

Proof: Since

$$f(x) - f(c) = \frac{f(x) - f(c)}{x - c} \cdot (x - c),$$

by Lemma 7.1 $\lim_{x \to c}(f(x) - f(c)) = f'(c) \cdot 0 = 0$, i.e. $f(x) \to f(c)$ as $x \to c$.   □

This result, though it may seem esoteric, is necessary in the standard results that follow:

Lemma 8.2   Suppose that f and g are differentiable at c. Then $\lambda f, f + g, fg$ and (provided $g(c) \neq 0$) f/g are all differentiable at c, their derivatives being, respectively, $\lambda f'(c)$, $f'(c) + g'(c)$, $f'(c)g(c) + f(c)g'(c)$, $(f'(c)g(c) - f(c)g'(c))/(g(c))^2$, where λ is a constant.

Proof: These are routine uses of Lemma 7.1, but notice the need for Theorem 8.1:

$$\frac{f(x)g(x) - f(c)g(c)}{x - c} = \frac{f(x) - f(c)}{x - c} g(x) + \frac{g(x) - g(c)}{x - c} f(c) \to f'(c)g(c) + g'(c)f(c)$$

where the continuity of g is used in the first part.   □

The consideration of differentiating $f \circ g$ leads to a problem whose solution stimulates an idea that turns out to have a wider usefulness. The natural impulse is to write $\{f(g(x)) - f(g(c))\}/\{x - c\}$ as the product of $\{f(g(x)) - f(g(c))\}/\{g(x) - g(c)\}$ and $\{g(x) - g(c)\}/\{x - c\}$, but this gives a meaningless expression if $g(x) = g(c)$ which may, of course, happen for values of x other than c. The way to avoid this difficulty is to notice that if f is differentiable at c then $f(x) - f(c) = (x - c)f'(c) + r(x)$ where $r(x)/(x - c) \to 0$ as $x \to c$, and that, conversely, if there is a constant α such that $f(x) = f(c) + \alpha(x - c) + r(x)$ where $r(x)/(x - c) \to 0$ as $x \to c$ then f is differentiable at c and $f'(c) = \alpha$. In this sense, f is differentiable if it is of the form of a linear function plus a 'small' correction.

Lemma 8.3   Let g be differentiable at a, and f be differentiable at $b = g(a)$. Then (assuming f and g are compatible so that $f \circ g$ exists) $f \circ g$ is differentiable at a and $(f \circ g)'(a) = f'(g(a)) \cdot g'(a)$.

Proof: Let $r_1(x) = g(x) - g(a) - (x - a)g'(a)$ and

$r_2(y) = f(y) - f(b) - (y-b)f'(b)$. Then $r_1(x)/(x-a) \to 0$ as $x \to a$ and $r_2(y)/(y-b) \to 0$ as $y \to b$. Also

$$(f \circ g)(x) - (f \circ g)(a) = (g(x) - g(a))f'(b) + r_2(g(x))$$
$$= (x-a)g'(a)f'(b) + f'(b)r_1(x) + r_2(g(x))$$

so if we set $r_3(x) = f'(b)r_1(x) + r_2(g(x))$ it only remains to show that $r_3(x)/(x-a) \to 0$ as $x \to a$. Since $r_1(x)/(x-a) \to 0$ as $x \to a$ we need only consider $r_2(g(x))/(x-a)$.

Let $\varepsilon > 0$. Since $r_2(y)/(y-b) \to 0$ as $y \to b$, $\exists \delta_1 > 0$ s.t.

$$0 < |y-b| < \delta_1 \Rightarrow |r_2(y)|/|y-b| < \varepsilon/(|g'(a)| + 1)$$
$$\Rightarrow |r_2(y)| < |y-b|\varepsilon/(|g'(a)| + 1).$$

Since $r_2(b) = 0$, we deduce that

$$|y-b| < \delta_1 \Rightarrow |r_2(y)| \leq |y-b|\varepsilon/(|g'(a)| + 1).$$

Now since g is differentiable, $\exists \delta_2 > 0$ such that

$$0 < |x-a| < \delta_2 \Rightarrow \left| \frac{g(x) - g(a)}{x-a} - g'(a) \right| < 1$$

$$\Rightarrow |g(x) - g(a)| < (|g'(a)| + 1)|x-a|.$$

Thus, if $\delta = \min(\delta_2, \delta_1/(|g'(a)| + 1))$,

$$0 < |x-a| < \delta \Rightarrow |g(x) - g(a)| < (|g'(a)| + 1)\delta \leq \delta_1$$
$$\Rightarrow |r_2(g(x))| \leq |g(x) - g(a)| \cdot \varepsilon/(|g'(a)| + 1) < \varepsilon |x-a|.$$

Since $\varepsilon > 0$ was arbitrary we have shown $r_2(g(x))/(x-a) \to 0$ as $x \to a$. $\square$

The Lemma just proved is called the chain rule, and is more expressively put by writing in Leibniz's notation:

$$\frac{d}{dx}(f(g(x))) = \frac{df}{dg} \cdot \frac{dg}{dx}.$$

The Newtonian notation makes it a little more explicit where the various functions are to be evaluated. It should be clear to the reader by now that the proof above was not obtained without a preliminary attempt, to see what the appropriate constants should be. For the sake of efficiency we shall clear one more piece of general theory out of the way before proceeding.

Lemma 8.4   Let $f:[a,b] \to \mathbf{R}$ be differentiable and possess an inverse function $g:[c,d] \to [a,b]$. Then, if $f'(x_0) \neq 0$, g is differentiable at $y_0 = f(x_0)$ and $g'(y_0) = 1/f'(x_0)$.

Proof: We know that g is continuous by Theorem 7.8. Also if $y \neq y_0$

$$\frac{g(y) - g(y_0)}{y - y_0} = \frac{g(y) - g(y_0)}{f(g(y)) - f(g(y_0))}.$$

Since f is differentiable and $f'(x_0) \neq 0$, $(x-x_0)/(f(x) - f(x_0)) \to 1/f'(x_0)$ as

$x \to x_0$, so

$$\forall \varepsilon > 0 \quad \exists \delta > 0 \quad \text{s.t.} \quad 0 < |x - x_0| < \delta \Rightarrow \left| \frac{x - x_0}{f(x) - f(x_0)} - \frac{1}{f'(x_0)} \right| < \varepsilon.$$

Then, by the continuity and injectivity of g, $\exists \delta_1 > 0$ s.t.

$$0 < |y - y_0| < \delta_1 \Rightarrow 0 < |g(y) - g(y_0)| < \delta \Rightarrow \left| \frac{g(y) - g(y_0)}{y - y_0} - \frac{1}{f'(x_0)} \right| < \varepsilon$$

(substituting $g(y)$ for x above). Thus g is differentiable at y_0 and $g'(y_0) = 1/f'(x_0)$. $\square$

The main result above is not the value of $g'(x_0)$, which we could calculate by the chain rule, but the knowledge that it exists. We may use this to show that the function $g:[0, \infty) \to [0, \infty)$ given by $g(x) = x^{1/n}$ is differentiable and $g'(x) = (1/n)x^{(1/n) - 1}$, since g is the inverse of $x \mapsto x^n$.

As is familiar from calculus, the derivative yields information about the function, although care is occasionally needed to distinguish those situations where we require to know the derivative at a range of points from those where the knowledge is only required at a single point.

Example Suppose that $f:[a, b] \to \mathbf{R}$ is continuous, and is differentiable at all points of (a, b). If $f(x_0)$ is the maximum value of $f(x)$ for $a \le x \le b$ then either x_0 is one of the endpoints or $f'(x_0) = 0$. The corresponding result for the minimum is also true of course.

Solution: Theorem 7.5 guarantees that there is an $x_0 \in [a, b]$ for which $f(x_0)$ is the maximum of $\{f(x) : a \le x \le b\}$. Suppose x_0 is not an endpoint so $a < x_0 < b$. Then if $x < x_0$, $(f(x) - f(x_0))/(x - x_0) \ge 0$ (numerator non-positive, since $f(x_0)$ is the maximum value). Thus,

$\lim_{x \to x_0 -} (f(x) - f(x_0))/(x - x_0) \ge 0$ so $f'(x_0) \ge 0$. For $x > x_0$, $(f(x) - f(x_0))/(x - x_0) \le 0$ and $f'(x_0) \le 0$, this time by considering the right-hand limit. Combining the inequalities, $f'(x_0) = 0$.

It is worth noticing that the maximum of $f(x)$ need not occur where $f'(x) = 0$, as is shown by considering $f(x) = x$ on $[0, 1]$. It is easy to overlook the possibility that the maximum may occur at an endpoint!

In deducing more information about a function from a knowledge of its derivatives we need some simple results which connect f with f' without involving limits explicitly. The principal theorem of this sort is the Mean Value Theorem, for which Rolle's Theorem is a preamble.

Theorem 8.5 (Rolle's Theorem) Suppose that $a < b$, that $f:[a, b] \to \mathbf{R}$ is continuous, that f is differentiable at all points of (a, b) and that $f(a) = f(b)$. Then there is a point $\xi \in (a, b)$ for which $f'(\xi) = 0$.

Proof: The idea is suggested by Fig. 8.2; ξ is the point at which f is maximum or minimum.

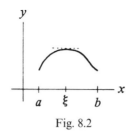

Fig. 8.2

By Theorem 7.5 there are points x_0, x_1 at which f attains its minimum and maximum, so $\forall x \in [a, b]$ $f(x_0) \leq f(x) \leq f(x_1)$. This is the key to the proof. If either x_0 or x_1 is interior to $[a, b]$ then f' is zero there by the example above, so we choose ξ to be x_0 or x_1, and $f'(\xi) = 0$. The only remaining possibility is that x_0 and x_1 are both endpoints of $[a, b]$. In this case, $f(x_0) = f(x_1)$ (since $f(a) = f(b)$) and f is constant on $[a, b]$ whence $f'(x) = 0$ for $a < x < b$ and we may choose $\xi = (a + b)/2$. $\square$

Theorem 8.6 (The Mean Value Theorem) Suppose that $a < b$, that $f : [a, b] \to \mathbf{R}$ is continuous and that f is differentiable at all points of (a, b). Then there is a point $\xi \in (a, b)$ for which

$$f'(\xi) = \frac{f(b) - f(a)}{b - a}.$$

Remarks: The number $(f(b) - f(a))/(b - a)$ is the average gradient of f between a and b, so the theorem states that at some point, f' takes its average (mean) value. (See Fig. 8.3.)

Proof: We reduce this to Rolle's Theorem (which is a special case). f satisfies all the hypotheses of Rolle's Theorem except that $f(a)$ and $f(b)$ may differ, so let $g(x) = f(x) + kx$ $(a \leq x \leq b)$ for k a constant. Then $g(a) = g(b)$ if we choose $k = -(f(b) - f(a))/(b - a)$, and the resulting function g is continuous and differentiable as required, so Rolle's Theorem guarantees the existence of $\xi \in (a, b)$ such that $g'(\xi) = 0$. Since $g'(\xi) = f'(\xi) + k$ we see $f'(\xi) = (f(b) - f(a))/(b - a)$. $\square$

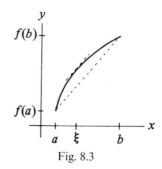

Fig. 8.3

The Mean Value Theorem is the simplest method of relating properties of f to those of f'. For example, if we know that for all $x \in (a, b)$, $f'(x) > 0$ then it follows that f is strictly increasing on $[a, b]$. To see this, choose $x, y \in [a, b]$ with $x < y$. Then applying the Mean Value Theorem to f on the interval $[x, y]$ shows that, for some $\xi \in (x, y)$, $f(y) - f(x) = (y - x)f'(\xi) > 0$. Thus $f(y) > f(x)$. This is a typical use of the Mean Value Theorem (MVT); some property of f' which is known to be true at all points of some interval is deployed to deduce information about f. We need to know about $f'(x)$ for all x in some interval since we do not usually have more precise details of the location of the 'ξ' of the theorem. Notice also that the 'a' and 'b' of the theorem may be chosen to be any two points in the interval on which the function is defined.

Less obviously, we can use the MVT's ideas to produce a very useful result about limits. The limit, as $x \to a$, of $f(x)/g(x)$ is $f(a)/g(a)$ provided both functions are continuous and $g(a) \neq 0$. It may happen, however, that $f(a)$ and $g(a)$ are both zero and the limit may exist even though our simple results on continuity do not show this; consider $\lim_{x \to 0} x/x$. 'L'Hôpital's Rule' (which was proved first by Bernoulli, not L'Hôpital) is a useful tool is this situation. We first need to modify the MVT.

Lemma 8.7 (Cauchy's Mean Value Theorem)   Let $a < b$, $f, g : [a, b] \to \mathbf{R}$ be two continuous functions, both differentiable at all points of (a, b), and suppose that $\forall x \in (a, b) \ g'(x) \neq 0$. Then there is a point $\xi \in (a, b)$ for which

$$\frac{f(b) - f(a)}{g(b) - g(a)} = \frac{f'(\xi)}{g'(\xi)}.$$

Proof: Let $\phi(x) = (g(b) - g(a))f(x) - (f(b) - f(a))g(x)$. Then ϕ is continuous on $[a, b]$, differentiable on (a, b) and $\phi(a) = \phi(b)$ so by Rolle's Theorem $\exists \xi \in (a, b)$ for which $\phi'(\xi) = 0$. Then, by assumption, $g'(\xi) \neq 0$ and $g(b) - g(a) \neq 0$ (since otherwise Rolle's Theorem applied to g would show $g'(x) = 0$ for some $x \in (a, b)$) so the result is obtained by dividing and rearranging the equation $\phi'(\xi) = 0$.   □

Theorem 8.8 (L'Hôpital's Rule)   Suppose that $f, g : [a, b] \to \mathbf{R}$ are continuous on $[a, b]$ and differentiable on (a, b) and that $f(a) = g(a) = 0$. Then if $f'(x)/g'(x)$ tends to a limit as $x \to a+$ so does $f(x)/g(x)$ and

$$\lim_{x \to a+} \frac{f(x)}{g(x)} = \lim_{x \to a+} \frac{f'(x)}{g'(x)}.$$

The corresponding results for left-hand and two-sided limits also hold.

Proof: Suppose $f'(x)/g'(x) \to L$ as $x \to a+$.
Let $\varepsilon > 0$.

$$\exists \delta > 0 \quad \text{s.t.} \quad 0 < x - a < \delta \Rightarrow |f'(x)/g'(x) - L| < \varepsilon. \qquad (1)$$

Now choose $x \in (a, a + \delta)$ and apply the Cauchy MVT to $[a, x]$. There is a number ξ with $a < \xi < x$ such that

$$\frac{f'(\xi)}{g'(\xi)} = \frac{f(x) - f(a)}{g(x) - g(a)} = \frac{f(x)}{g(x)}.$$

Thus

$$\left| \frac{f(x)}{g(x)} - L \right| = \left| \frac{f'(\xi)}{g'(\xi)} - L \right| < \varepsilon \quad \text{by (1)},$$

since $0 < \xi - a < x - a < \delta$. Since x was a typical point of $(a, a + \delta)$ we have proved that $\forall x \in (a, a + \delta) \, | f(x)/g(x) - L | < \varepsilon$ and, noticing now that $\varepsilon > 0$ was arbitrary, we have shown that $\lim_{x \to a+} f(x)/g(x) = L$.

Example Find

$$\lim_{x \to 0} \frac{\sqrt{(1 + x)} - x/2 - 1}{x^2}.$$

Both numerator and denominator are differentiable and have the value 0 at $x = 0$, so by L'Hôpital's Rule, our limit exists if

$$\lim_{x \to 0} \frac{1/(2\sqrt{(1 + x)}) - 1/2}{2x} = \lim_{x \to 0} \frac{1/\sqrt{(1 + x)} - 1}{4x}$$

exists. This limit is equally indeterminate as it stands, so we apply L'Hôpital again and the second limit will exist if

$$\lim_{x \to 0} \frac{(-1/2)(1 + x)^{-3/2}}{4}$$

exists. This limit clearly does exist and has the value $-1/8$, so the two previous limits are shown to exist in turn and have the value $-1/8$.

In the statements of the Mean Value Theorem and its near relatives, we assumed $a < b$. This is not an issue if we allocate the labels a and b to two numbers of our choice since a may be chosen to be the smaller, but it may be a nuisance if a and b have predetermined meanings. Notice, however, that the result of the MVT does not rely on the order of a and b in that $(f(b) - f(a))/(b - a)$ is unaltered if we interchange the roles of a and b. For this reason we may relax our requirement that $a < b$ to one that $a \neq b$ provided we alter the statement about ξ to be that ξ lie strictly between a and b; the word 'strict' denotes that ξ is not allowed to be equal to either a or b. The same comment applies to Taylor's Theorem below, the last elaboration of the MVT.

Theorem 8.9 (Taylor's Theorem) Suppose $a \neq b$, that f is defined on the closed interval defined by a and b and that f possesses derivatives of order up to n at all points of the interval. Then

$$f(b) = f(a) + (b - a)f'(a) + \frac{(b - a)^2}{2!} f''(a) + \cdots + \frac{(b - a)^{n-1}}{(n - 1)!} f^{(n-1)}(a) + R_n$$

where, for some point ξ between a and b,

$$R_n = \frac{(b - a)^n}{n!} f^{(n)}(\xi).$$

Proof: We shall apply the MVT to a suitable function, some care being needed to choose one which does not give the intermediate derivatives in the remainder term. Define $\phi:[a,b]\to\mathbf{R}$ by

$$\phi(x) = f(x) + (b-x)f'(x) + \cdots + \frac{(b-x)^{n-1}}{(n-1)!}f^{(n-1)}(x).$$

ϕ is continuous and differentiable on $[a,b]$, since each part of it is, and $\phi'(x) = (b-x)^{n-1}f^{(n)}(x)/(n-1)!$, so, to dispose of the $(b-\xi)^{n-1}$ term which would otherwise occur, we let $\psi(x) = (b-x)^n$ and apply Cauchy's MVT, yielding the existence of a point ξ for which

$$\frac{R_n}{-(b-a)^n} = \frac{\phi(b)-\phi(a)}{\psi(b)-\psi(a)} = \frac{\phi'(\xi)}{\psi'(\xi)} = \frac{(b-\xi)^{n-1}f^{(n)}(\xi)}{-(n-1)!n(b-\xi)^{n-1}}.$$

Hence

$$R_n = \frac{(b-a)^n}{n!}f^{(n)}(\xi),$$

as promised. $\square$

The whole value of Taylor's Theorem is in the expression for the 'remainder' R_n. If we suppose f satisfies the hypotheses of the theorem, then if $a < x \le b$ the restriction of f to $[a,x]$ is n times differentiable and applying the theorem shows that

$$f(x) = f(a) + (x-a)f'(a) + \cdots + \frac{(x-a)^{n-1}}{(n-1)!}f^{(n-1)}(a) + R_n$$

where, for some ξ_x between a and x,

$$R_n = (x-a)^n f^{(n)}(\xi_x)/n!.$$

Thus f may be approximated by the polynomial expression on the right, the error in this approximation being R_n. Whether this is useful or not depends on the size of R_n, of course. For 'ordinary' functions we have much choice as to the value of n which we select, and it is usual to choose n as small as possible for the purposes in hand; we shall see this in examples.

Had we used the ordinary MVT instead of Cauchy's the expression obtained for R_n would have been $(b-a)(b-\xi)^{n-1}f^{(n)}(\xi)/(n-1)!$, which is not as useful as the one we obtained above.

Example Suppose that f is n times differentiable in $[a-h,a+h]$ for some $h>0$ and that $f'(a)=f''(a)=\cdots=f^{(n-1)}(a)=0\neq f^{(n)}(a)$. Then if $f^{(n)}$ is continuous,

f has a local maximum at a if n is even and $f^{(n)}(a)<0$,

f has a local minimum at a if n is even and $f^{(n)}(a)>0$,

f has a point of inflexion at a if n is odd.

Solution: Here we use the remainder of order n since $f^{(n)}(a)$ is the lowest order derivative which is non-zero at a. Let $x\in[a-h,a+h]$, so by Taylor,

$$f(x) = f(a) + \frac{(x-a)^n}{n!}f^{(n)}(\xi)$$

where ξ is some point, depending on x, between a and x. Since $f^{(n)}$ is continuous, there is an interval $(a - \delta, a + \delta)$ such that if ξ belongs to this interval, $f^{(n)}(\xi)$ has the same sign as $f^{(n)}(a)$. Therefore, for all $x \in [a - \delta, a + \delta]$, $f(x) - f(a) = (x - a)^n f^{(n)}(\xi)/n!$ has the same sign as $(x - a)^n f^{(n)}(a)$. If n is even, so $(x - a)^n \geq 0$, $f(x) - f(a)$ has constant sign for $x \in (a - \delta, a + \delta)$ and if $f^{(n)}(a) > 0$ this shows $f(x) \geq f(a)$ so there is a local minimum at a, while $f^{(n)}(a) < 0$ indicates a local maximum. If n is odd, $(x - a)^n$ is positive for $a < x \leq a + \delta$ and negative for $a - \delta \leq x < a$ so $f(x) - f(a)$ changes sign in the interval $(a - \delta, a + \delta)$, no matter how small δ is, so f has a point of inflexion at a.

This example is typical of many uses of Taylor's Theorem in that we need to use the remainder of the correct order; the hypotheses tell us nothing directly about R_{n-1} or, even if f is $(n + 1)$-times differentiable, R_{n+1}.

We can use our knowledge of the behaviour of functions to simplify the treatment of sequences. Earlier we considered sequences of the form $a_1 = \alpha$, $a_{n+1} = f(a_n)$ where f is a given function. Some of our work there can be simplified. We first need a lemma:

Lemma 8.10   Suppose that f is continuous at a and that for each $n \in \mathbf{N}$ a_n belongs to the domain of f. Then if $a_n \to a$ as $n \to \infty$, $f(a_n) \to f(a)$ as $n \to \infty$.

Proof: Let $\varepsilon > 0$. Since f is continuous at a, $\exists \delta > 0$ s.t. $|x - a| < \delta \Rightarrow |f(x) - f(a)| < \varepsilon$. Since $a_n \to a$ as $n \to \infty$, $\exists N$ s.t. $\forall n \geq N\ |a_n - a| < \delta$. Therefore $\forall n \geq N\ |f(a_n) - f(a)| < \varepsilon$.

Since $\varepsilon > 0$ was arbitrary, we have shown that $\forall \varepsilon > 0\ \exists N$ s.t. $\forall n \geq N$ $|f(a_n) - f(a)| < \varepsilon$, as required. $\square$

From this it follows that the limit, a, of the sequence defined by $a_{n+1} = f(a_n)$, if there is one, will satisfy $a = f(a)$. Also, we may investigate the increasing or decreasing nature of (a_n) by noticing that $a_{n+1} - a_n = f(a_n) - f(a_{n-1}) = (a_n - a_{n-1})f'(\xi_n)$ for some ξ_n between a_n and a_{n-1}, by the MVT.

Example   Let $a_1 = 1$, $a_{n+1} = \frac{1}{2}(a_n + 2/a_n)$. Then if we set $f(x) = \frac{1}{2}(x + 2/x)$ we see that f is differentiable on $(0, \infty)$ and $f'(x) = \frac{1}{2}(1 - 2/x^2)$.

It is obvious that $x > 0 \Rightarrow f(x) > 0$, so by induction, $a_n > 0$ for all n since $a_n > 0 \Rightarrow a_{n+1} = f(a_n) > 0$ and $a_1 > 0$. Sketching the graph of f shows that it has a minimum at $x = \sqrt{2}$ and, omitting the details, we notice that $x > 0 \Rightarrow f(x) \geq \sqrt{2}$. From this we see that $\forall n \geq 1\ a_{n+1} = f(a_n) \geq \sqrt{2}$, thus $\forall n \geq 2\ a_n \geq \sqrt{2}$. (See Fig. 8.4.)

Also $a_{n+1} - a_n = f'(\xi_n)(a_n - a_{n-1})$ so, since ξ_n lies between a_{n-1} and a_n, we see that $\forall n \geq 3$, $\xi_n \geq \sqrt{2}$ so $f'(\xi_n) \geq 0$ and $a_{n+1} - a_n$ has the same sign as $a_n - a_{n-1}$, hence the same sign as $a_3 - a_2$. Since $a_2 = 3/2$ and $a_3 = 17/12$, we see that after the second term, (a_n) decreases and is bounded below by 0, so there is an a s.t. $a_n \to a$ as $n \to \infty$. Since a satisfies $a = f(a)$, $a^2 = 2$ so $a_n \to \sqrt{2}$ as $n \to \infty$.

This may also be seen by noticing from the graph that if $x \geq \sqrt{2}$, $f(x) \leq x$, hence $a_n \geq \sqrt{2} \Rightarrow a_{n+1} \leq a_n$. Thus, plotting a_n on the x-axis, if we mark the intercept of the line $x = a_n$ with the graph of f and draw the horizontal line

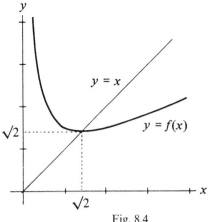

Fig. 8.4

from this point $(a_n, f(a_n))$ to the line $y = x$ we obtain the point $(f(a_n), f(a_n))$, i.e. (a_{n+1}, a_{n+1}). Dropping the vertical line to the x-axis shows the position of a_{n+1}. This method is very helpful but care must be taken to draw the graph sufficiently accurately: it is important to notice whether, say, a minimum lies to the left or the right of the intersection of the graph of f and the line $y = x$. This is easier to illustrate in the next example.

Example Let $a_1 = \alpha$, $a_{n+1} = \mu a_n(1 - a_n)$ where $0 \le \alpha \le 1$ and $0 < \mu \le 2$. Then, letting $f(x) = \mu x(1 - x)$, we see that $f(x) = x$ iff $x = 0$ or $x = 1 - 1/\mu$, so the only possible limits are 0 and $1 - 1/\mu$. f has a maximum at $x = 1/2$, with $f(1/2) = \mu/4$. It is easy to show by induction that $\forall n \in \mathbf{N}\ 0 \le a_n \le 1$.

Case 1: $0 < \mu \le 1$. Here the only solution of $x = f(x)$ which lies in $[0, 1]$ is 0, so this is the only potential limit; we prove $a_n \to 0$ as $n \to \infty$. Since $a_{n+1} - a_n = (\mu - 1)a_n - \mu a_n^2$, if $\mu \le 1$ (a_n) is decreasing; it is also bounded below by 0. Filling in the details, we see that $a_n \to 0$. (See Fig. 8.5.)

Case 2: $1 < \mu \le 2$. Since the maximum value of f occurs at $x = 1/2$, we see that $0 \le x \le 1 - 1/\mu \Rightarrow f'(x) \ge 0$ so f is increasing on $[0, 1 - 1/\mu]$. Therefore,

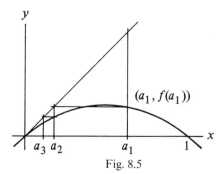

Fig. 8.5

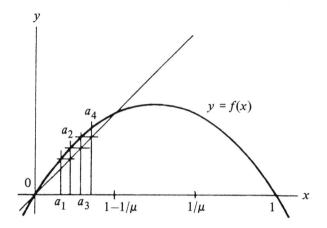

Fig. 8.6

$0 \le x \le 1 - 1/\mu \Rightarrow 0 = f(0) \le f(x) \le f(1 - 1/\mu) = 1 - 1/\mu$, so $a_n \in [0, 1 - 1/\mu] \Rightarrow$
$a_{n+1} \in [0, 1 - 1/\mu]$. Thus if $0 \le \alpha \le 1 - 1/\mu$, (a_n) is increasing and
bounded above; if $\alpha = 0$, $a_n \to 0$ otherwise the limit cannot be 0 (it must
be at least a_1) whence it is $1 - 1/\mu$. If $1/\mu \le \alpha \le 1$, $0 \le a_2 \le 1 - 1/\mu$, so the
sequence tends to 0 if $a_2 = 0$ (i.e. $\alpha = 1$) and to $1 - 1/\mu$ otherwise. This leaves
the case where $1 - 1/\mu < \alpha < 1/\mu$, and Fig. 8.6 shows that in this event, $\forall n \in \mathbf{N}$,
$1 - 1/\mu < a_n < 1/\mu$ and (a_n) decreases, and is bounded below. Here $a_n \to 1 - 1/\mu$
(since $a_n \to 0$ is impossible, and 0 and $1 - 1/\mu$ are the only two potential limits).
The implication $1 - 1/\mu < x < 1/\mu \Rightarrow 1 - 1/\mu < f(x) < \mu/4 \le 1/\mu$ supplies the
main part of the last argument. Readers should fill in all the details to convince
themselves that everything above is valid.

　　Had we chosen $\mu > 2$ in the above example, we would have the maximum
of f occurring at $x = 1/2$, between the points where $f(x) = x$. This means
that $x \in (0, 1 - 1/\mu) \nRightarrow f(x) \in (0, 1 - 1/\mu)$ and (a_n) need not be increasing. Not all
is lost, however, as we may use the MVT to show that the distance of a_n
from its limit tends to zero. Let $a = 1 - 1/\mu$ so that $f(a) = a$. Then

$$|a_{n+1} - a| = |f(a_n) - f(a)| = |f'(\xi_n)| |a_n - a| \qquad (*)$$

where ξ_n is between a_n and a. Provided we can show that for all points ξ in
an interval containing all the a_n, $|f'(\xi)| \le \gamma$ where γ is a constant less than
1, we will be able to obtain a useful result.

　　From Fig. 8.7, it is clear that $0 < a_n \le 1 - 1/\mu \Rightarrow a_{n+1} \ge a_n$, whence we
see that if $0 < \alpha \le 1 - 1/\mu$, the first few terms of the sequence will increase
in value. Since $f(x)$ has a maximum at $x = 1/2$ and this maximum value is
$\mu/4$, it is also obvious that $\forall n \ge 2$ $a_n = f(a_{n-1}) \le \mu/4$.

　　For definiteness, let us choose $\mu = 5/2$. Then let $a = 3/5$ so that $f(a) = a$.
Since $f(\frac{1}{2}) = 5/8$ we see that $\forall n \ge 2$ $a_n \le 5/8$. Moreover,
$f(5/8) = 75/128 > 1/2$, so, since f is decreasing on the interval $[1/2, 5/8]$,
we see that $1/2 \le x \le 5/8 \Rightarrow f(5/8) \le f(x) \le f(1/2) \Rightarrow 1/2 \le f(x) \le 5/8$. Thus
$a_n \in [1/2, 5/8] \Rightarrow a_{n+1} \in [1/2, 5/8]$.

　　If $\alpha = 0$ or 1 then $a_n = 0$ $\forall n \ge 2$ and $a_n \to 0$ as $n \to \infty$. If $0 < a_n \le 3/5$ then

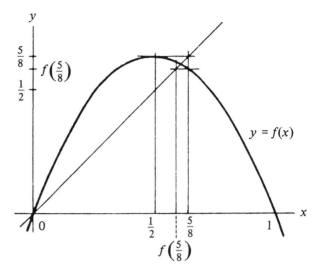

Fig. 8.7. The function $f(x) = \mu x(1-x)$ for $\mu = 5/2$; in this case $1 - 1/\mu = 3/5$.

$a_{n+1} \geq a_n$, so if $0 < \alpha \leq 3/5$ it follows that either (a_n) is increasing or there is a value N such that $a_N > 3/5$. Since $\forall n \geq 2$ $a_n \leq 5/8$, the first case implies that (a_n) is increasing and bounded above, hence (a_n) tends to some limit. This limit will satisfy $f(x) = x$, hence is either 0 or 3/5, and since the limit cannot be less than α (since $\forall n \in \mathbf{N}$ $a_n \geq a_1 = \alpha$), $a_n \to 3/5$ as $n \to \infty$. In the second case we see that $\forall n \geq N$ $a_n \in [1/2, 5/8]$. Then

$$|a_{n+1} - a| = |f(a_n) - f(a)| = |f'(\xi_n)| |a_n - a| \qquad (*)$$

for some ξ_n between a and a_n, hence in $[1/2, 5/8]$. Since $f'(x) = 5/2 - 5x$ we see that $|f'(x)| \leq |f'(5/8)| = 5/8$ for all $x \in [1/2, 5/8]$. From $(*)$ then, $\forall n \geq N$ $|a_{n+1} - a| \leq (5/8)|a_n - a|$, so $\forall n \geq N$ $|a_n - a| \leq (5/8)^{n-N}|a_N - a|$, whence $a_n \to a$. The cases for $\alpha > 3/5$ can be seen immediately on noticing that $0 \leq a_2 < 3/5$.

The method just used will tackle values of μ in the range $2 < \mu < 1 + \sqrt{3}$. Considerably more cunning is needed to deal with $1 + \sqrt{3} \leq \mu \leq 3$. For values of μ between 3 and 4, the sequence does not usually tend to a limit; consider the case $\mu = 7/2$ and $\alpha = 3/7$. The behaviour for such μ is very complicated and difficult to determine and great care is needed in inferring results from diagrams. (Some idea of this can be gained by testing the values on a computer.)

In fact, the behaviour of this sequence, given by $a_{n+1} = \mu a_n(1 - a_n)$, and others like it, is not yet fully understood and this sort of problem is an active field of research. For $1 < \mu \leq 3$ we saw that, apart from the cases where $\alpha = 0$ or 1, $a_n \to a = 1 - 1/\mu$, this being the non-zero solution of the equation $x = f(x)$ which a limit of (a_n) must satisfy. For $\mu > 3$, $|f'(a)| > 1$ and there is an interval $(a - \delta, a + \delta)$ on which $|f'|$ is greater than 1. By the MVT, in the form $|a_{n+1} - a| = |f'(\xi_n)| |a_n - a|$, we see that if a_n and a_{n+1} both lie in $(a - \delta, a + \delta)$ then $|f'(\xi_n)| > 1$ and, unless $a_n = a$, a_{n+1} is further from a than

a_n is. The sequence is 'repelled' from a and, apart from exceptional cases where there is a value of N with $a_N = a$, the sequence cannot tend to a. For μ slightly greater than 3 it turns out that alternate terms tend to two distinct limits, the sequence of even-numbered terms tending to p, say, while the odd-numbered terms tend to q. In this case $f(p) = q$ and $f(q) = p$, both numbers satisfying the equation $x = f(f(x))$. The limit a could be said to have 'split' into two, alternate terms of the sequence tending to these two. For larger μ the two points p and q split into four, then eight, sixteen and so on, in these cases the sequence (a_n), for large n, being nearly periodic, the values being close to a number of points in turn. For still larger values of μ the behaviour becomes chaotic and the sequence need have little discernible pattern, while the behaviour appears to vary unpredictably with changes in the value of α. For a readable account of this, see May (1976).

Another example occurs with Newton's method for finding the root of an equation $f(x) = 0$. Suppose that $a < b$ and that f is twice differentiable on $[a, b]$, that $f(a)$ and $f(b)$ have opposite signs and that both f' and f'' are non-zero and have constant sign on $[a, b]$. Let (a_n) be obtained by choosing a_1 to be whichever of a or b has the property that f and f'' have the same sign, and letting $a_{n+1} = a_n - f(a_n)/f'(a_n)$. Then (a_n) tends to the root of $f(x) = 0$ lying between a and b.

(Some relaxation of the conditions may be allowed, but at the expense of some complication, and perhaps doubt in determining which solution of $f(x) = 0$ is the limit.)

We shall tackle the case illustrated in Fig. 8.8, where $f(a) < 0 < f(b)$ and $f'(x) > 0$, $f''(x) > 0$ on $[a, b]$. By the Intermediate Value Theorem (IVT) there is at least one solution of the equation $f(x) = 0$ between a and b, and the positivity of f' ensures that there is no more than one: call it x_0. The diagram suggests that $\forall n \in \mathbf{N} \; x_0 \le a_n \le b$ and that (a_n) decreases.

Let $\phi(x) = x - f(x)/f'(x)$ so that $a_{n+1} = \phi(a_n)$ and

$$x > x_0 \Rightarrow \phi'(x) = f(x)f''(x)/(f'(x))^2 > 0.$$

Then $\qquad a_n > x_0 \Rightarrow \phi(a_n) - \phi(x_0) = (a_n - x_0)\phi'(\xi_n) > 0 \Rightarrow a_{n+1} > x_0$

(where ξ_n is some point between x_0 and a_n). Also if $x_0 \le a_n \le b$ then

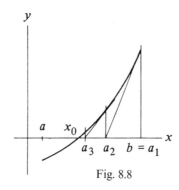

Fig. 8.8

$a_{n+1} = a_n - f(a_n)/f'(a_n) \leq a_n$. Induction now shows that (a_n) decreases and we see that $a_n \to x_0$.

The convergence here is eventually 'rapid' in that

$$a_{n+1} - x_0 = \frac{(a_n - x_0)^2}{2} \phi''(\xi_n)$$

for some ξ_n between x_0 and a_n (by Taylor's Theorem applied to ϕ) so that if $|\phi''(x)| \leq M$ for $x \in [a, b]$, $|a_{n+1} - x_0| \leq \frac{1}{2}M|a_n - x_0|^2$ and if a_n is very close to x_0, a_{n+1} will be considerably closer. A precise evaluation of the consequences of this is best left to a course on numerical mathematics.

As a final example of the use of Taylor's Theorem we shall consider a few finite series. In common with most examples of Taylor's Theorem, we use the number of terms in the theorem appropriate to our problem. The motivation is the set of induction results giving

$$\sum_{k=1}^{n} k = \frac{1}{2}n(n+1), \quad \sum_{k=1}^{n} k^2 = n(n+1)(2n+1)/6, \quad \sum_{k=1}^{n} k^3 = n^2(n+1)^2/4$$

and so on. In all these, the coefficient of the highest power of n in $\sum_{k=1}^{n} k^{\alpha}$ is $1/(\alpha + 1)$; we shall show this is general.

Let $f(x) = x^{\alpha+1}/(\alpha + 1)$ so that f has derivatives of all orders. By Taylor's Theorem, for $k \in \mathbf{N}$ and $\alpha \geq 1$,

$$((k+1)^{\alpha+1} - k^{\alpha+1})/(\alpha + 1) = k^{\alpha} + \frac{1}{2}\alpha\xi_k^{\alpha-1}$$

for some $\xi_k \in (k, k+1)$, so, summing from $k = 1$ to $n - 1$,

$$(n^{\alpha+1} - 1)/(\alpha + 1) = \sum_{k=1}^{n-1} k^{\alpha} + R_n$$

where

$$R_n = \frac{1}{2}\alpha \sum_{k=1}^{n-1} \xi_k^{\alpha-1} \leq \frac{1}{2}\alpha \sum_{k=1}^{n-1} (k+1)^{\alpha-1} \leq \frac{1}{2}\alpha \cdot n \cdot n^{\alpha-1}.$$

Thus

$$\sum_{k=1}^{n} k^{\alpha} = n^{\alpha} + (n^{\alpha+1} - 1)/(\alpha + 1) - R_n$$

$$= n^{\alpha+1}/(\alpha + 1) + R'_n$$

where $R'_n = n^{\alpha} - 1/(\alpha + 1) - R_n$ so that $|R'_n| \leq (1 + \frac{1}{2}\alpha)n^{\alpha} \; \forall n \in \mathbf{N}$. Therefore the leading term is $n^{\alpha+1}/(\alpha + 1)$ (for $R'_n/n^{\alpha+1} \to 0$ as $n \to \infty$).

This rather neatly illustrates that one gets out of Taylor's Theorem what one puts in. If we take an extra term, we obtain

$$((k+1)^{\alpha+1} - k^{\alpha+1})/(\alpha + 1) = k^{\alpha} + \frac{1}{2}\alpha k^{\alpha-1} + \alpha(\alpha - 1)\eta_k^{\alpha-2}/6$$

where $\eta_k \in (k, k+1)$. Then summing from $k = 1$ to $n - 1$ gives

$$(n^{\alpha+1} - 1)/(\alpha + 1) = \sum_{k=1}^{n-1} k^{\alpha} + \frac{1}{2}\alpha \sum_{k=1}^{n-1} k^{\alpha-1} + R_n$$

where here $|R_n| \leq \alpha(\alpha - 1)n^{\alpha-1}/6$. Adjusting, and noticing that

$$\sum_{k=1}^{n-1} k^{\alpha-1} = -n^{\alpha-1} + \sum_{k=1}^{n} k^{\alpha-1} = -n^{\alpha-1} + n^{\alpha}/\alpha + S_n$$

where
$$|S_n| \leq Mn^{\alpha - 1}$$

(for constant M), we obtain

$$\sum_{k=1}^{n} k^{\alpha} = n^{\alpha + 1}/(\alpha + 1) + \tfrac{1}{2}n^{\alpha} + T_n$$

where, for some constant M', $\forall n \in \mathbf{N}$ $|T_n| \leq M'n^{\alpha - 1}$ $(\alpha \geq 2)$. Notice the curiosity that the coefficient of n^{α} does not depend on α. This method can, of course, be taken further.

Problems

1. Let $f(x) = |x| \, \forall x \in \mathbf{R}$. Show that f is differentiable at all points of $\mathbf{R}$ except 0 and find its derivative.

2. Suppose that $f_1, \ldots, f_n$ are differentiable and that, for $i = 1, 2, \ldots, n$, $f_i(c) \neq 0$. Let $h(x) = f_1(x)f_2(x)$ and show that $h'(c)/h(c) = f'_1(c)/f_1(c) + f'_2(c)/f_2(c)$. Show also that if $k(x) = f_1(x) \ldots f_n(x)$ then $k'(c)/k(c) = \sum_{i=1}^{n} f'_i(c)/f_i(c)$.

3. Let f, g and h be differentiable. Calculate the derivatives of the functions whose value at x is $f(2x), f(x^2), f(g(x) \cdot h(x)), f(g(h(x)))$.

4. Let $f:[a, b] \rightarrow \mathbf{R}$ be continuous, and be differentiable on (a, b). Use the MVT to show that if $\forall x \in (a, b)$ $f'(x) = 0$, then f is constant.

5. Suppose that $f:\mathbf{R} \rightarrow \mathbf{R}$ is differentiable and that $\forall x \in \mathbf{R}$ $|f'(x)| \leq M$. Prove that $\forall x, y \in \mathbf{R}$ $|f(x) - f(y)| \leq M|x - y|$.

6. Suppose that $g:\mathbf{R} \rightarrow \mathbf{R}$ and that $\forall x, y \in \mathbf{R}$ $|g(x) - g(y)| \leq M|x - y|^2$, where M is a constant. Prove that g is constant.

7. In each case below, the function described has a stationary value at $x = 0$; decide whether it is a maximum, minimum or point of inflexion: x^7, $\cos x - 1 + x^2/2$, $x^2/(1 + x)^2$, $x^2 - \sin^2 x$, $\sin x - \tan x$.

8. Let f be twice differentiable and suppose that $f(a) = f(b) = f(c)$, where $a < b < c$. Prove that there is a point ξ between a and c for which $f''(\xi) = 0$.

9. Show that if f is strictly increasing and differentiable on (a, b) then $\forall x \in (a, b)$ $f'(x) \geq 0$. Give an example to show that $f'(x)$ may be zero for some points x.

10. (i) Let $a < b < c$ and suppose that $\phi:(a, c) \rightarrow \mathbf{R}$ is differentiable. Let $\phi'(b) > 0$. Deduce that $\exists \delta > 0$ s.t. $0 < |x - b| < \delta \Rightarrow \phi(x) - \phi(b)$ and $x - b$ have the same sign. Hence show that $b < x < b + \delta \Rightarrow \phi(x) > \phi(b)$ and $b - \delta < x < b \Rightarrow \phi(x) < \phi(b)$.

    (ii) Let $a < b < c, f:(a, c) \rightarrow \mathbf{R}$ be twice differentiable, $f'(b) = 0$ and $f''(b) > 0$. Show that $\exists \delta > 0$ s.t. $b < x < b + \delta \Rightarrow f'(x) > 0$ and hence that, for the same $\delta, b < x < b + \delta \Rightarrow f(x) > f(b)$. Deduce that if f has a local maximum at b, $f''(b) \leq 0$. (Unlike the result proved in the text, we do not assume f'' is continuous here.)

11. The function $y:[a,b] \to \mathbf{R}$ is continuous, and twice differentiable at all points of (a,b). It is also known that $y(a) = y(b) = 0$. By considering the point at which y attains its maximum or minimum, prove that if y is not identically zero, either there is a $\xi \in (a,b)$ for which $y''(\xi) \le 0$ and $y(\xi) > 0$ or there is a $\xi \in (a,b)$ with $y''(\xi) \ge 0$ and $y(\xi) < 0$. (Use Q10.)

    If, in addition, y satisfies the differential equation

    $$y''(x) + f(x)y'(x) - g(x)y(x) = 0 \quad (\forall x \in (a,b))$$

    where $\forall x \in (a,b)\ g(x) > 0$, show y is identically zero.

12. Evaluate the limits as $x \to 0$ of:

    $$\frac{\sqrt{(1+x)} - 1}{\sqrt{(1-x)} - 1}, \quad \frac{x^2 - x}{\sqrt{(1+x)} - 1}, \quad \frac{\sin x}{x}, \quad \frac{\sqrt{(1+x^2)} - 1}{x \sin x}.$$

13. Define (a_n) by $a_1 = 1$, $a_{n+1} = 1 + 1/a_n$. By considering the function f given by $f(x) = 1 + 1/x$ and showing $3/2 \le x \le 2 \Rightarrow 3/2 \le f(x) \le 2$, or otherwise, show that $a_n \to (1 + \sqrt{5})/2$ as $n \to \infty$. Investigate the limit when different choices are made for $a_1 > 0$.

14. Let $a_1 = \alpha$ and $\forall n \in \mathbf{N}\ a_{n+1} = 1 + 1/(1 + a_n)$. Show that if $\alpha \ge 0$ then $\forall n \ge 2\ a_n \ge 1$, and $|a_{n+1} - \sqrt{2}| \le (1/4)|a_n - \sqrt{2}|$. Deduce that $a_n \to \sqrt{2}$ as $n \to \infty$.

15. Let $f(x) = 2\sqrt{x}\ (x \ge 0)$. By applying Taylor's Theorem to f, show that for $n \in \mathbf{N}\ 2\sqrt{(n+1)} - 2\sqrt{n} = 1/\sqrt{n} - 1/(4\xi_n^{3/2})$ for some $\xi_n \in (n, n+1)$. By showing that $\sum \xi_n^{-3/2}$ converges, deduce that if $S_n = 1 + 1/\sqrt{2} + \cdots + 1/\sqrt{n}$, $S_n - 2\sqrt{(n+1)}$ tends to a limit as $n \to \infty$.

16. Assuming the required properties of the sine function, define f by $f(0) = 0$, $f(x) = x^2 \sin(1/x)\ (x \ne 0)$, and show that $f'(0) = 0$. Observe that f' is not continuous at 0.

17.* Suppose that $f:(a,b) \to \mathbf{R}$ is differentiable. Show that f' has the intermediate value property even if the derivative is discontinuous, as follows: First show that if $f:[c,d] \to \mathbf{R}$ is not injective then $\exists \xi \in [c,d]$ s.t. $f'(\xi) = 0$. Deduce that if $\forall x \in (c,d)\ f'(x) \ne 0$, then f is strictly increasing or strictly decreasing on $[c,d]$. Now suppose that $f'(c) < 0 < f'(d)$ and deduce the existence of $\xi \in (c,d)$ with $f'(\xi) = 0$, where c and d are two typical points of (a,b).

CHAPTER 9

Functions Defined by Power Series

We are led naturally to the study of those functions which can be represented as the sums of power series for two reasons. Taylor's Theorem allows us to express a function f as a sum of finitely many terms of the form $(x - a)^n f^{(n)}(a)/n!$ plus a remainder, and we may observe that in some cases the remainder tends to zero as the number of terms increases; this may be thought of as expressing f as a sum of the simpler functions $(x - a)^n$. More significantly, we arrive at power series by attempting to find functions with desirable properties, usually arising from differential equations. If we wish to find a function equal to its own derivative, then if this function were expressible in the form $\sum_{n=0}^{\infty} a_n x^n$, and if the derivative turned out to equal the expression obtained by differentiating term by term, $\sum_{n=1}^{\infty} n a_n x^{n-1}$, then the equality of the function and its derivative would be guaranteed if we were to choose the coefficients a_n so that these two series were identical, that is, $a_n = 1/n!$. Assuming this outline programme is correct and, in particular, that the interchange of the limits involved in taking derivatives of infinite sums can be justified, this yields a simple technique for producing various special functions required in everyday mathematics.

The first thing to recall is the result of Lemma 5.11, that if the power series $\sum a_n w^n$ converges, then $\sum a_n x^n$ converges if $|x| < |w|$. We systematise this:

Definition The power series $\sum a_n x^n$ is said to have **radius of convergence** R if it converges for all real x with $|x| < R$ and diverges for all real x with $|x| > R$. (Notice that nothing is said about $x = \pm R$.)

Theorem 9.1 A power series $\sum a_n x^n$ either converges for all $x \in \mathbf{R}$ or has a radius of convergence.

Proof: We introduce the set $A = \{x \geq 0 : \sum a_n x^n \text{ converges}\}$. The two cases arise according as A is bounded above or not.

Suppose A is bounded above. Since A is non-empty ($0 \in A$), it has a supremum; call it R. We show that R is the radius of convergence.

Let $-R < x < R$. Since $|x| < \sup A$, $\exists w \in A$ with $|x| < w$. Then since $\sum a_n w^n$ converges, Lemma 5.11 shows that $\sum a_n x^n$ converges absolutely. Thus, for all $x \in (-R, R)$ $\sum a_n x^n$ converges.

Now let $x \in \mathbf{R}$ with $|x| > R$. Then $\sum a_n x^n$ must diverge, for if not, let $y = (|x| + R)/2$ so that $|y| < |x|$ and by Lemma 5.11 $\sum a_n y^n$ would converge, giving $y \in A$ and $y > \sup A = R$, a contradiction.

For the remaining case, suppose A is not bounded above. Let $x \in \mathbf{R}$. Then $|x|$ is not an upper bound for A, so $\exists w \in A$ such that $w > |x|$, whence Lemma 5.11 proves that $\sum a_n x^n$ converges. Since x was arbitrary, we have shown that the series converges for all $x \in \mathbf{R}$. $\square$

Note: The proof above actually shows more than we stated: the series is *absolutely* convergent for $|x| < R$ in the first case and for all $x \in \mathbf{R}$ in the second.

Examples The ratio test shows that $\sum x^n$ has radius of convergence 1 and that $\sum x^n/n!$ converges (absolutely) for all real x.

Theorem 9.2 Let the power series $\sum a_n x^n$ converge absolutely for all $x \in (-R, R)$ and define

$$f(x) = \sum_{n=0}^{\infty} a_n x^n \quad (-R < x < R).$$

f is continuous.

Proof: Caution is needed here. The individual functions $a_n x^n$ summed in the series are all continuous, but we take a limit in forming the sum to infinity, so we need to *prove* f continuous and avoid interchanging the order of limits in this proof. The crux of the proof is the reduction to a finite sum.

Let $x_0 \in (-R, R)$; we prove that f is continuous at x_0.

Let $\varepsilon > 0$. Choose r with $|x_0| < r < R$, which is certainly possible. Since $\sum |a_n r^n|$ converges, there is a N such that

$$\sum_{n=N+1}^{\infty} |a_n r^n| = \left| \sum_{n=0}^{\infty} |a_n r^n| - \sum_{n=0}^{N} |a_n r^n| \right| < \varepsilon/3. \tag{1}$$

Now the function $x \mapsto \sum_{n=0}^{N} a_n x^n$ is the sum of $N + 1$ continuous functions and is therefore continuous, so there is a $\delta_1 > 0$ for which

$$|x - x_0| < \delta_1 \Rightarrow \left| \sum_{n=0}^{N} a_n x^n - \sum_{n=0}^{N} a_n x_0^n \right| < \varepsilon/3. \tag{2}$$

We now need to consider only x satisfying $|x| \leq r$ and $|x - x_0| < \delta_1$, so let $\delta = \min(\delta_1, r - |x_0|)$. Then $|x - x_0| < \delta \Rightarrow |x| < r$ and $|x - x_0| < \delta_1$ so that, if $|x - x_0| < \delta$,

$$|f(x) - f(x_0)| = \left| \sum_{n=0}^{\infty} a_n x^n - \sum_{n=0}^{\infty} a_n x_0^n \right|$$

$$\leq \left| \sum_{n=N+1}^{\infty} a_n x^n \right| + \left| \sum_{n=0}^{N} a_n x^n - \sum_{n=0}^{N} a_n x_0^n \right| + \left| \sum_{n=N+1}^{\infty} a_n x_0^n \right|$$

$$< \sum_{n=N+1}^{\infty} |a_n x^n| + \varepsilon/3 + \sum_{n=N+1}^{\infty} |a_n x_0^n| \qquad \text{by (2)}$$

$$\leq 2 \sum_{n=N+1}^{\infty} |a_n| r^n + \varepsilon/3 \quad (\text{since} \quad |x| \leq r, \quad |x_0| \leq r)$$

$$< \varepsilon \qquad\qquad\qquad\qquad\qquad\qquad\qquad\qquad \text{by (1).}$$

Thus $\forall \varepsilon > 0 \, \exists \delta > 0$ s.t. $|x - x_0| < \delta \Rightarrow |f(x) - f(x_0)| < \varepsilon$, and f is continuous at x_0. $\quad\square$

Note: The key step is the observation that for all x with $|x| \leq r$ $\sum_{n=N+1}^{\infty} |a_n x^n| \leq \sum_{n=N+1}^{\infty} |a_n r^n| < \varepsilon/3$, that is, we have a single value of N which ensures this inequality for all of these values of x. The result of the theorem is that, in these specific circumstances, we may interchange the limits

$$\lim_{x \to x_0} \sum_{n=0}^{\infty} a_n x^n = \sum_{n=0}^{\infty} \lim_{x \to x_0} (a_n x^n).$$

The continuity of sums of power series is the key to the following 'identity theorem' allowing us to equate the terms of two power series. It is the continuity which allows us to deduce the step we need for $x = 0$ when the preliminary information was obtained subject to the restriction $x \neq 0$.

Theorem 9.3   Let $\sum a_n x^n$ and $\sum b_n x^n$ be two power series and let $f(x) = \sum_{n=0}^{\infty} a_n x^n$ and $g(x) = \sum_{n=0}^{\infty} b_n x^n$, each valid when the appropriate series converges. If there is an $R > 0$ such that $\forall x \in (-R, R)$ $f(x) = g(x)$, the series are identical, i.e. $\forall n \geq 0$ $a_n = b_n$.

Proof: Suppose that $\forall x \in (-R, R)$ $f(x) = g(x)$, where $R > 0$. Then setting $x = 0$ shows that $a_0 = b_0$.
Now suppose that $a_n = b_n$ for $n = 0, 1, \ldots, N$. Then, for $x \in (-R, R)$,

$$0 = f(x) - g(x) = \sum_{n=0}^{\infty} (a_n - b_n) x^n = \sum_{n=N+1}^{\infty} (a_n - b_n) x^n = x^{N+1} h(x)$$

where $h(x) = \sum_{n=0}^{\infty} (a_{n+N+1} - b_{n+N+1}) x^n$. Clearly $h(x) = 0$ if $x \in (-R, R) \setminus \{0\}$. Since h is the sum of a power series, convergent for $|x| < R$ (and thus absolutely convergent), h is continuous, so $h(0) = \lim_{x \to 0} h(x) = 0$. Substituting shows $a_{N+1} = b_{N+1}$.
It follows by induction that $\forall n \geq 0$ $a_n = b_n$. $\quad\square$
The next step in our development of a calculus for power series is to consider differentiation, in particular, to consider the series obtained by differentiating each term of the power series and that which yields the original series when we differentiate the terms of the new series.

Theorem 9.4   The series $\sum a_n x^n$, $\sum n a_n x^{n-1}$ and $\sum (a_n/(n+1)) x^{n+1}$ either all have the same radius of convergence or all converge for all $x \in \mathbf{R}$.

Proof: Since all the series converge for $x = 0$, we need only consider $x \neq 0$.
Suppose that $\sum n a_n x^{n-1}$ converges absolutely for all $x \in (-R_2, R_2)$. Let $b_n = n a_n / x$ so that $n a_n x^{n-1} = b_n x^n$. Since $|a_n| \leq |n a_n| = |b_n x|$ for $n \geq 1$, $\sum |a_n x^n|$ is convergent for $x \in (-R_2, R_2)$ by comparison with $\sum |x| \cdot |b_n x^n|$.

Now suppose, instead, that $\sum a_n x^n$ is absolutely convergent for all $x \in (-R_1, R_1)$. Let $0 < |x| < R_1$ and choose w such that $|x| < |w| < R_1$. Then $\sum |a_n w^n|$ is convergent. Since $|w/x| > 1$, Bernoulli's Inequality shows that $\forall n \in \mathbb{N} \; |w/x|^n > n(|w/x| - 1)$ whence $|w|^n > n|x|^{n-1}(|w| - |x|)$ so $|n a_n x^{n-1}| < |a_n w^n|/(|w| - |x|)$. By comparison with $\sum |a_n w^n|$, we see that $\sum |n a_n x^{n-1}|$ converges. Since x was typical, we deduce that $\sum |n a_n x^{n-1}|$ converges for all $x \in (-R_1, R_1)$.

The two paragraphs above show that if either of the series $\sum a_n x^n$ and $\sum n a_n x^{n-1}$ converges for all $x \in \mathbb{R}$ then so does the other. (Suppose the first series converges for all $x \in \mathbb{R}$; then R_1 may be chosen as large as we wish, and, given $x_0 \in \mathbb{R}$, we may choose $R_1 > |x_0|$ to deduce that the second series converges at $x = x_0$.) The remaining case is where both series have a radius of convergence, say R_1 and R_2 respectively. Then by the first paragraph, $R_1 \geq R_2$ (since $\sum a_n x^n$ converges for $|x| < R_2$) while the second paragraph shows $R_2 \geq R_1$.

The result for $\sum (a_n/(n+1)) x^{n+1}$ is obtained from the above by setting $c_0 = 0$, $c_n = a_{n-1}/n$ ($n \geq 1$) and considering $\sum c_n x^n$ and $\sum n c_n x^{n-1}$.   □

Theorem 9.5   Suppose that for all $x \in (-R, R)$ the series $\sum a_n x^n$ converges absolutely and that $f(x) = \sum_{n=0}^{\infty} a_n x^n$. Then f possesses derivatives of all orders, $f'(x) = \sum_{n=1}^{\infty} n a_n x^{n-1}$ and the kth derivative $f^{(k)}(x) = \sum_{n=k}^{\infty} n(n-1) \ldots (n-k+1) a_n x^{n-k}$.

Proof: We shall prove the result for f'. The general case then follows by observing that f' is given by the sum of a power series so the theorem applies to f', and so on.

Choose $x \in (-R, R)$ and r satisfying $|x| < r < R$. These we shall keep fixed. Considering the function $y \mapsto y^n$ we see that there is a ξ_n between x and y for which

$$y^n = x^n + (y - x) n x^{n-1} + \tfrac{1}{2}(y - x)^2 n(n-1) \xi_n^{n-2}.$$

ξ_n here depends on n and y, but for all $y \in (-r, r)$, $|\xi_n| < r$ so, since all the series converge by comparison with $\sum |a_n r^n|$, $\sum |n a_n r^{n-1}|$ and $\sum n(n-1)|a_n r^{n-2}|$, which converge by Theorem 9.4,

$$f(y) - f(x) = \sum_{n=0}^{\infty} a_n(y^n - x^n) = \sum_{n=1}^{\infty} a_n(y^n - x^n)$$

$$= \sum_{n=1}^{\infty} n a_n(y - x) x^{n-1} + \sum_{n=1}^{\infty} \tfrac{1}{2} n(n-1) a_n(y - x)^2 \xi_n^{n-2}.$$

Thus

$$\left| \frac{f(y) - f(x)}{y - x} - \sum_{n=1}^{\infty} n a_n x^{n-1} \right| \leq |y - x| \cdot \sum_{n=1}^{\infty} \tfrac{1}{2} n(n-1)|a_n \xi_n^{n-2}|$$

$$\leq M|y - x|, \qquad (1)$$

where $M = \sum_{n=1}^{\infty} \tfrac{1}{2} n(n-1)|a_n r^{n-2}|$, a constant independent of y. It is now clear that $\lim_{y \to x} (f(y) - f(x))/(y - x) = \sum_{n=1}^{\infty} n a_n x^{n-1}$.   □

Again, notice that the introduction of r allowed us to obtain the estimate with the number M independent of y, and thus (1) is valid for all $y \in (x - \delta, x + \delta)$ where $\delta = r - |x|$. The result is another in which the interchange of two limiting operations (summing to infinity and differentiating) is justified.

Suppose we seek a function f which is equal to its own derivative. If we presume that the function can be expressed in the form $f(x) = \sum_{n=0}^{\infty} a_n x^n$, where the series converges for $x \in (- R, R)$, then $f'(x) = \sum_{n=1}^{\infty} n a_n x^{n-1}$ so $f' = f$ implies that $\forall x \in (- R, R) \sum_{n=0}^{\infty} a_n x^n = \sum_{n=1}^{\infty} n a_n x^{n-1} = \sum_{n=0}^{\infty} b_n x^n$ where $b_n = (n + 1)a_{n+1}$. By Theorem 9.3, then, $\forall n \geq 0$, $a_n = b_n = (n + 1)a_{n+1}$. Thus $a_n = a_0/n!$ and $f(x) = a_0 \sum_{n=0}^{\infty} x^n/n!$. At this stage we know that if there is a function satisfying the equation $f' = f$ and given by the power series, then it is of the form given; we still have to check that there is such a function. This is now easy, for we can check that the series converges for all $x \in \mathbf{R}$ and so the calculations are all justified in retrospect.

Definition We define the **exponential function** exp: $\mathbf{R} \to \mathbf{R}$ by

$$\exp(x) = \sum_{n=0}^{\infty} x^n/n!.$$

We now know that the exponential function is continuous, differentiable as often as we wish, and that it equals its own derivative. To go further we shall use a general theorem whose proof requires a good deal of stamina and organisation. The aim is to consider the result of multiplying two power series $(\sum_{n=0}^{\infty} a_n x^n) \cdot (\sum_{n=0}^{\infty} b_n x^n)$. Ignoring, for the moment, all qualms about justification, we may hope this product can be expressed as a power series in x, and that the coefficient of x^n in it is obtained by collecting the terms from the two brackets, $a_r x^r$ being matched with $b_{n-r} x^{n-r}$; the expected coefficient is then $a_0 b_n + \cdots + a_n b_0$. This result is true, though we shall only prove it where the series are absolutely convergent. It may be false if neither series converges absolutely.

Theorem 9.6 (The Cauchy Product Theorem) Let x be a number for which the two power series $\sum a_n x^n$ and $\sum b_n x^n$ are both absolutely convergent. Define $c_n = \sum_{r=0}^{n} a_r b_{n-r}$. Then the series $\sum c_n x^n$ is absolutely convergent and

$$\sum_{n=0}^{\infty} c_n x^n = \left(\sum_{n=0}^{\infty} a_n x^n \right) \cdot \left(\sum_{n=0}^{\infty} b_n x^n \right).$$

Proof: We first need to identify the nature of our task. Since

$$c_k = a_0 b_k + a_1 b_{k-1} + \cdots + a_k b_0,$$

$$c_k x^k = \sum_{i=0}^{k} (a_i x^i)(b_{k-i} x^{k-i})$$

$$= \sum \{(a_i x^i)(b_j x^j): \; i, j \geq 0, \; i + j = k\}.$$

(In what follows, i, j, k and n will always denote integers.) Therefore, $c_k x^k$ is

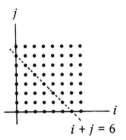

Fig. 9.1

the sum of all those terms of the form $(a_i x^i)(b_j x^j)$ where the pair of integers (i, j), plotted on a graph, lies on that portion of the line $i + j = k$ in the first quadrant ($i \geq 0$, $j \geq 0$ ensuring this). We now see that $\sum_{n=0}^{N} c_n x^n$ equals the sum of all terms $(a_i x^i)(b_j x^j)$ where $i + j$ takes in turn the values $0, 1, \ldots, N$, that is, the sum of all the terms whose indices (i, j) lie in the triangular region defined by $i \geq 0$, $j \geq 0$, $i + j \leq N$. (See Fig. 9.1.) $(\sum_{i=0}^{\infty} a_i x^i)(\sum_{j=0}^{\infty} b_j x^j)$ is the limit as $N \to \infty$ of $(\sum_{i=0}^{N} a_i x^i)(\sum_{j=0}^{N} b_j x^j)$. Now the latter expression is the sum of all the terms of the form $(a_i x^i)(b_j x^j)$ where the indices lie in the square region $0 \leq i, j \leq N$.

Let $s_n = \sum_{i=0}^{n} a_i x^i$, $t_n = \sum_{j=0}^{n} b_j x^j$ and $u_n = \sum_{k=0}^{n} c_k x^k$ for $n = 0, 1, \ldots$. From the discussion above, $s_N t_N$ will not generally be equal to u_M for any M. The technique is to use the absolute convergence of $\sum a_i x^i$ and $\sum b_j x^j$ to prove first that $\sum c_k x^k$ is absolutely convergent, so that by choosing N and n large enough, $s_N t_N$ and u_n will be close to their respective limits. If we at the same time ensure that $n \geq 2N$ the terms $(a_i x^i)(b_j x^j)$ summed in $s_N t_N$ will all contribute to the sum forming u_n, so $u_n - s_N t_N$ will consist of the sum of those terms $(a_i x^i)(b_j x^j)$ where (i, j) lies in the region outside the small square but inside the triangle in Fig. 9.2. It is here that we use the absolute convergence: because the sum of all terms $|(a_i x^i)(b_j x^j)|$ which have one index greater than N can be shown to be small, the same is true of our chosen selection of these terms since the terms omitted, being moduli, are non-negative. Let us start.

Let $s'_n = \sum_{i=0}^{n} |a_i x^i|$, $t'_n = \sum_{j=0}^{n} |b_j x^j|$ and $u'_n = \sum_{k=0}^{n} |c_k x^k|$. Since $\sum |a_i x^i|$ and $\sum |b_j x^j|$ converge there are numbers s, s', t, t' such that $s_n \to s$, $s'_n \to s'$, $t_n \to t$ and $t'_n \to t'$ as $n \to \infty$. Moreover, since (s'_n) and (t'_n) are increasing, $\forall n \in \mathbf{N}$ $s'_n \leq s'$ and $t'_n \leq t'$.

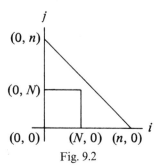

Fig. 9.2

For each $k \in \mathbf{N}$, the triangle inequality shows that

$$|c_k x^k| = |\sum \{(a_i x^i)(b_j x^j): \quad i \geq 0, \quad j \geq 0, \quad i+j=k\}|$$
$$\leq \sum \{|a_i x^i| \cdot |b_j x^j|: \quad i \geq 0, \quad j \geq 0, \quad i+j=k\}.$$

Hence, for all n,

$$u'_n = \sum_{k=0}^{n} |c_k x^k| \leq \sum_{k=0}^{n} (\sum \{|a_i x^i| \cdot |b_j x^j|: \quad i \geq 0, \quad j \geq 0, \quad i+j=k\})$$
$$= \sum \{|a_i x^i| \cdot |b_j x^j|: \quad i \geq 0, \quad j \geq 0, \quad i+j \leq n\}$$
$$\leq \sum \{|a_i x^i| \cdot |b_j x^j|: \quad 0 \leq i \leq n, \quad 0 \leq j \leq n\}$$
$$= s'_n t'_n \leq s' t'.$$

Since this is true for all $n \in \mathbf{N}$, (u'_n) is bounded above and hence $\sum |c_k x^k|$ converges.

To find the sum of $\sum c_k x^k$ we need more effort. Notice first that we may presume s' and t' are both positive since if, say, $s' = 0$ then $\forall n \geq 0$, $a_n = 0$ and the whole result is trivial.

Let $\varepsilon > 0$. Then there exist N_1, N_2 and N_3 such that

$$\left. \begin{array}{l} \forall n \geq N_1 \quad |s'_n - s'| < \varepsilon/(4t'), \\[2pt] \forall n \geq N_2 \quad |t'_n - t'| < \varepsilon/(4s'), \\[6pt] \forall n \geq N_3 \quad |s_n t_n - st| < \varepsilon/2. \end{array} \right\} \qquad (1)$$

and

Let $N = \max(N_1, N_2, N_3)$ so that all three inequalities hold if $n \geq N$. Recalling that if $n \geq 2N$, all the terms $(a_i x^i)(b_j x^j)$ which occur in $s_N t_N$ occur in u_n (see Fig. 9.2), we have

$$\forall n \geq 2N \quad u_n = \sum \{(a_i x^i)(b_j x^j): \quad i,j \geq 0, \quad i+j \leq n\}$$
$$= \sum \{(a_i x^i)(b_j x^j): \quad 0 \leq i,j \leq N\}$$
$$+ \sum \{(a_i x^i)(b_j x^j): \quad i,j \geq 0, \quad i+j \leq n, \quad \max(i,j) > N\}.$$

Since the first sum is equal to $s_N t_N$, we have $\forall n \geq 2N$

$$|u_n - s_N t_N| \leq \sum \{|a_i x^i| \cdot |b_j x^j|: \quad i,j \geq 0, \quad i+j \leq n, \quad \max(i,j) > N\} \qquad ⓦ$$
$$\leq \sum \{|a_i x^i| \cdot |b_j x^j|: \quad 0 \leq i,j \leq n, \qquad \max(i,j) > N\} \qquad ⓧ$$
$$= \sum \{|a_i x^i| \cdot |b_j x^j|: \quad 0 \leq i \leq n, \quad N < j \leq n\} \qquad ⓨ$$
$$+ \sum \{|a_i x^i| \cdot |b_j x^j|: \quad N < i \leq n, \quad 0 \leq j \leq N\} \qquad ⓩ$$
$$= s'_n(t'_n - t'_N) + (s'_n - s'_N)t'_N. \qquad (2)$$

To see this, the sets of indices (i,j) occurring in sums ⓦ, ⓧ, ⓨ and ⓩ are indicated in Fig. 9.3.

Finally,

$$\forall n \geq 2N \quad |u_n - st| \leq |u_n - s_N t_N| + |s_N t_N - st|$$
$$< s'_n(t'_n - t'_N) + (s'_n - s'_N)t'_N + \tfrac{1}{2}\varepsilon \quad \text{(by (1) and (2))}$$
$$\leq s'(t' - t'_N) + (s' - s'_N)t' + \tfrac{1}{2}\varepsilon$$
$$< s' \cdot \varepsilon/(4s') + t' \cdot \varepsilon/(4t') + \tfrac{1}{2}\varepsilon = \varepsilon \quad \text{(by (1))}.$$

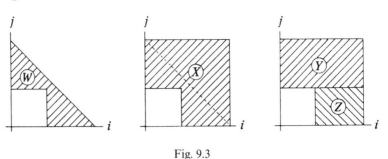

Fig. 9.3

Since $\varepsilon > 0$ was arbitrary we have shown that

$$\forall \varepsilon > 0 \quad \exists N \quad \text{s.t.} \quad \forall n \geq 2N \quad |u_n - st| < \varepsilon,$$

so $u_n \to st$ as $n \to \infty$, as required. $\square$

Remarks: The result need not be about power series, just the corresponding 'product' result for absolutely convergent series. This is obtained by setting $x = 1$ above.

This allows us to produce the required properties of the exponential function:

Theorem 9.7   (i) The exponential function is continuous, possesses derivatives of all orders and $(d/dx)(\exp x) = \exp x$. (ii) $\forall x, y \in \mathbf{R}$ $\exp(x + y) = (\exp x) \cdot (\exp y)$. (iii) $\exp(0) = 1$ and $\forall x \in \mathbf{R}$ $\exp x > 0$.

Proof: (i) is immediate from Theorems 9.2 and 9.5.

(ii) Since the series for $\exp x$ and $\exp y$ converge absolutely we can apply the Cauchy Product Theorem. Let $a_n = x^n/n!$ and $b_n = y^n/n!$ so that $\sum a_n$ and $\sum b_n$ are absolutely convergent. It follows that

$$(\exp x)(\exp y) = \left(\sum_{n=0}^{\infty} a_n \right)\left(\sum_{n=0}^{\infty} b_n \right) = \sum_{n=0}^{\infty} c_n$$

where

$$c_n = \sum_{i=0}^{n} a_i b_{n-i} = \sum_{i=0}^{n} x^i y^{n-i}/(i!(n-i)!)$$

$$= \sum_{i=0}^{n} \binom{n}{i} x^i y^{n-i}/n! = (x + y)^n/n!.$$

(iii) The value of $\exp(0)$ is obvious. Also it is clear that $x > 0 \Rightarrow \exp x = 1 + \sum_{n=1}^{\infty} x^n/n! > 1$. By parts (i) and (ii), $(\exp x)(\exp(-x)) = 1$, whence $x < 0 \Rightarrow \exp x = 1/\exp(-x) > 0$. $\square$

We define the number e by $\exp(1)$. Then since $1 + 1 = 2$, it follows that $\exp(2) = e^2$ and it is easily checked that if n is an integer $\exp(n) = e^n$, while a little more work will check this for rational powers of e. We therefore *define* e^x for irrational powers x by the equation $e^x = \exp(x)$, noticing that this coincides with our old definition for rational x. The number e is about

2.71828...; the series converges so rapidly that it is easy to evaluate it to a high accuracy. (See Problems.)

The results of Theorem 9.7 are well known, but the following growth properties are equally useful—and their proof illustrates some simple but effective techniques.

Theorem 9.8 (i) The exponential function is strictly increasing, $e^x \to \infty$ as $x \to \infty$, and $e^x \to 0$ as $x \to -\infty$, (ii) $\{e^x : x \in \mathbf{R}\} = (0, \infty)$ and (iii) for all integers k, $e^x/x^k \to \infty$ as $x \to \infty$ and $x^k e^{-x} \to 0$ as $x \to \infty$.

Proof: (i) That exp is strictly increasing is immediate on noticing that its derivative is positive at all points. Now, for $x > 0$, $e^x = 1 + x + \sum_{n=2}^{\infty} x^n/n! > x$, so $\forall R \in \mathbf{R} \ \exists X$ (e.g. $X = |R|$) s.t. $x \geq X \Rightarrow e^x > R$. It follows that $e^x \to \infty$ as $x \to \infty$. The remaining part is clear on noticing that

$$\forall \varepsilon > 0 \quad x < -1/\varepsilon \Rightarrow e^{-x} > e^{1/\varepsilon} > 1/\varepsilon \Rightarrow e^x = 1/e^{-x} < \varepsilon.$$

(ii) We know that $x \in \mathbf{R} \Rightarrow e^x \in (0, \infty)$. Let $y \in (0, \infty)$. Then by the properties of part (i), $\exists x_0$ and $x_1 \in \mathbf{R}$ s.t. $e^{x_0} < y$ and $e^{x_1} > y$. The Intermediate Value Theorem now shows $\exists x \in \mathbf{R}$ s.t. $e^x = y$. This proves (ii).

(iii) Let k be an integer. If $k < 0$ it is clear that $e^x/x^k \to \infty$ (since both factors tend to ∞), so we may suppose $k \geq 0$. For $x > 0$, $e^x = \sum_{n=0}^{\infty} x^n/n! > x^{k+1}/(k+1)!$, so $e^x/x^k > x/(k+1)!$ Therefore $\forall R > 0 \ \exists X$ (e.g. $X = (k+1)!R$) s.t. $x \geq X \Rightarrow e^x/x^k > R$ showing that $e^x/x^k \to \infty$ as $x \to \infty$. The final part is now easily deduced from the observation that $x^k e^{-x}$ is the reciprocal of e^x/x^k. □

The intuitive result from Theorem 9.8 is that in a product of an exponential and a power, the exponential dominates the behaviour.

Since the exponential function is strictly increasing and has image $(0, \infty)$, it has an inverse function.

Definition The **logarithm function** log: $(0, \infty) \to \mathbf{R}$ is defined to be the inverse function of exp.

Theorem 9.9 The logarithm function is continuous, strictly increasing and differentiable and:

(i) $(d/dx)(\log x) = 1/x$, (ii) $\forall x, y > 0 \ \log(xy) = \log x + \log y$,
(iii) $\log 1 = 0$ and $\log x > 0 \Leftrightarrow x > 1$,
(iv) $\log x \to \infty$ as $x \to \infty$ but for all positive integers k, $x^{-k} \log x \to 0$ as $x \to \infty$,
(v) $\log x \to -\infty$ as $x \to 0+$ but for all positive integers k, $x^k \log x \to 0$ as $x \to 0+$.

Proof: (i) The differentiability arises from Lemma 8.4 on inverse functions, which also shows that log is strictly increasing. This result, or the chain rule, now allows the value of the derivative to be verified, since $x = e^{\log x} \ \forall x \in \mathbf{R}$.
(ii) follows on noticing that $\forall x, y \in \mathbf{R} \ xy = (e^{\log x})(e^{\log y}) = e^{\log x + \log y}$. (iii) is immediate since $\exp 0 = 1$ and log is strictly increasing. (iv) Since exp is strictly increasing, $\log x > R \Leftrightarrow x = e^{\log x} > e^R$ which immediately shows that $\log x \to \infty$ as $x \to \infty$.

As $x^{-k} \leq x^{-1}$ for $k \geq 1$ and $x \geq 1$, to show that $x^{-k} \log x \to 0$ as $x \to \infty$, it is enough to prove that $x^{-1} \log x \to 0$. Let $\varepsilon > 0$. We know already that $y e^{-y} \to 0$ as $y \to \infty$, so $\exists Y$ s.t. $y \geq Y \Rightarrow |y e^{-y}| < \varepsilon$. Then $x \geq e^{Y} \Rightarrow \log x \geq Y$ $\Rightarrow |x^{-1} \log x| = |\log x \cdot e^{-\log x}| < \varepsilon$. We have shown $x^{-1} \log x \to 0$ as $x \to \infty$. (v) is obtained from (iv) by using the fact that $x \to 0+$ if and only if $1/x \to \infty$. By (iv), since $\log y \to \infty$ as $y \to \infty$, $\forall R' \in \mathbf{R} \ \exists Y$ s.t. $\forall y \geq Y \log y > R'$. Let $R \in \mathbf{R}$. Then $\exists Y$ s.t. $\forall y \geq Y \ \log y > -R$, whence $0 < x \leq 1/Y \Rightarrow 1/x \geq Y \Rightarrow \log x < R$. We have shown $\log x \to -\infty$ as $x \to 0+$. $\lim\limits_{x \to 0+} x^k \log x$ is obtained from (iv) in a similar way. $\square$

We use the log function to define irrational powers of real numbers. Let $a > 0$ and $q \in \mathbf{Q}$. Then, since $a = e^{\log a}$, it is easily checked that $a^q = e^{q \log a}$. We extend this:

Definition If $a > 0$ and $x \in \mathbf{R}$ we define $a^x = e^{x \log a}$.

Notice that this coincides with the old definition of a^x for rational x, but that it is now clear that the function $x \mapsto a^x$ is differentiable, and its derivative can be evaluated.

Before we show what else can be done with our techniques, let us prove the General Binomial Theorem. The proof introduces an idea at least as useful as the result.

Theorem 9.10 (The General Binomial Theorem)   For $\alpha \in \mathbf{R}$ let

$$\binom{\alpha}{0} = 1 \quad \text{and} \quad \binom{\alpha}{n} = \frac{\alpha(\alpha - 1) \ldots (\alpha - n + 1)}{n!}.$$

Then, provided $|x| < 1$,

$$(1 + x)^\alpha = \sum_{n=0}^{\infty} \binom{\alpha}{n} x^n.$$

Proof: The ratio test shows that the series has radius of convergence 1, so define $f:(-1, 1) \to \mathbf{R}$ by $f(x) = \sum_{n=0}^{\infty} \binom{\alpha}{n} x^n$. Since f is given by a power series,

$f'(x) = \sum_{n=1}^{\infty} \binom{\alpha}{n} n x^{n-1}$. By observing that

$$(n + 1)\binom{\alpha}{n+1} + n\binom{\alpha}{n} = \alpha \binom{\alpha}{n}$$

it can be verified that $(1 + x)f'(x) = \alpha f(x)$. In solving this differential equation we see that

$$(d/dx)((1 + x)^{-\alpha} f(x)) = -\alpha(1 + x)^{-\alpha - 1} f(x) + (1 + x)^{-\alpha} f'(x)$$
$$= 0$$

so that $(1 + x)^{-\alpha} f(x)$ is a constant. Substituting $x = 0$ shows that the constant is 1, yielding the result. We leave the details to the reader. $\square$

The technique just used, of defining a function by means of a power series and then finding its properties by observing that it satisfies a simple differential equation, is quite useful. It helps, of course, when we have a good idea what properties we are looking for! In other cases, different features of the series give rise to the properties involved; in the case of the sine and cosine functions we cannot readily use differential equations since the solution of these equations requires a knowledge of the functions we seek. We can, however, use power series to obtain solutions of differential equations where no other technique is available.

Example   Suppose we wish to solve the differential equation:

$$x^2 \frac{d^2 y}{dx^2} + x \frac{dy}{dx} + (x^2 - 1)y(x) = 0.$$

(The equation is a particular case of Bessel's equation, and it arises frequently in certain applied mathematics. For details of this, see Dunning-Davies (1982), Chapter 13.)

If we assume, temporarily, that $y = \sum_{n=0}^{\infty} a_n x^n$ and that this series converges for all $x \in (-R, R)$ where $R > 0$, then by applying our theorems on power series we obtain

$$\sum_{n=2}^{\infty} n(n-1)a_n x^n + \sum_{n=1}^{\infty} na_n x^n + \sum_{n=0}^{\infty} a_n(x^{n+2} - x^n) = 0.$$

A little care is needed for low values of n here, since the lowest power of x in the first series is 2, in the second 1, and so on. By writing the x^0 and x^1 terms separately and collecting powers of x^n we deduce that

$$-a_0 x^0 + \sum_{m=2}^{\infty} ((m^2 - 1)a_m + a_{m-2})x^m = 0.$$

Since this is true for all $x \in (-R, R)$, we may use Theorem 9.3 to deduce that the coefficients are all zero, that is, $a_0 = 0$ and $\forall m \geq 2 \ (m^2 - 1)a_m + a_{m-2} = 0$. These two equations show us immediately that $a_m = 0$ for all even m, while, for odd m, $a_m = -a_{m-2}/(m^2 - 1)$. Thus $a_3 = -a_1/(4.2)$, $a_5 = a_1/(6.4.4.2)$ and we may check by induction that $\forall n \in \mathbb{N} \ a_{2n+1} = (-1)^n a_1/(2^{2n} \cdot n!(n+1)!)$. Therefore

$$y(x) = a_1 \sum_{n=0}^{\infty} \left(-\frac{1}{4}\right)^n \frac{x^{2n+1}}{n!(n+1)!}.$$

The ratio test will confirm that this series converges for all $x \in \mathbb{R}$, which gives us the information we need to check the validity of the whole process. We define a function $J_1(x)$ by

$$J_1(x) = \sum_{n=0}^{\infty} (-1)^n \frac{(x/2)^{2n+1}}{n!(n+1)!} \qquad (x \in \mathbb{R})$$

and check, via the arithmetic above, that it satisfies the differential equation. The determination of the properties of J_1, and a whole sequence of similar functions, is quite involved so we shall not pursue the matter. (For more information, see Dunning-Davies (1982).) Notice, however, that we have only produced one solution of the differential equation, where we would expect two independent solutions. Close examination of the logic shows that if there is a second independent solution, it cannot be of the form of a function given by a power series.

Examples Consider the function $f: \mathbf{R} \to \mathbf{R}$ defined by $f(x) = \exp(-1/x^2)$ $(x \neq 0)$, $f(0) = 0$. Since $e^{-y} \to 0$ as $y \to \infty$, we see that if $\varepsilon > 0$ $\exists Y$ s.t. $y \geq Y \Rightarrow |e^{-y}| < \varepsilon$ whence $0 < |x| \leq 1/\sqrt{Y} \Rightarrow |f(x)| < \varepsilon$. Therefore f is continuous at 0.

Since $\sqrt{y} e^{-y} \to 0$ as $y \to \infty$ (Problem 3), we see that if $\varepsilon > 0$ $\exists Y$ s.t. $y \geq Y \Rightarrow |\sqrt{y} e^{-y}| < \varepsilon$, whence $0 < |x| < 1/\sqrt{Y} \Rightarrow |f(x)/x| < \varepsilon$. This shows f is differentiable at 0 and $f'(0) = 0$; at all non-zero x, $f'(x) = (2/x^3) \exp(-1/x^2)$.

Pursuing these ideas shows that f possesses derivatives of all orders at 0 and that $\forall n \in \mathbf{N}$ $f^{(n)}(0) = 0$. If f were expressible as a power series $\sum a_n x^n$ for $x \in (-R, R)$ then we see by Theorem 9.5 that $a_n = f^{(n)}(0)/n! = 0$. In this case the series $\sum f^{(n)}(0) x^n/n!$ converges for all $x \in \mathbf{R}$ but for $x \neq 0$ its sum is not $f(x)$. The mere existence of all the derivatives of f is no guarantee that $\sum f^{(n)}(0) x^n/n!$ converges to $f(x)$. (This may also be seen by calculating the remainder R_n in Taylor's Theorem; in this case $R_n(x) = f(x)$ for all n, so $R_n(x) \not\to 0$.) This also shows that even if two functions, g and h, and all their derivatives are equal at one point a, the two functions need not be equal at any other point; consider $g(x) = h(x) + \exp(-1/x^2)$.

As a final theme, we observe that we can very easily obtain an estimate of the difference between the sum of the first N terms and the sum to infinity of a power series.

Theorem 9.11 Suppose that the power series $\sum a_n x^n$ converges absolutely for all $x \in (-R, R)$, and that $0 < r < R$. Then, for each $N \in \mathbf{N}$, there is a constant K such that

$$\forall x \in [-r, r] \quad \left| \sum_{n=0}^{\infty} a_n x^n - \sum_{n=0}^{N-1} a_n x^n \right| \leq K |x|^N.$$

Proof:

$$\left| \sum_{n=0}^{\infty} a_n x^n - \sum_{n=0}^{N-1} a_n x^n \right| = \left| \sum_{n=N}^{\infty} a_n x^n \right| \leq |x|^N \sum_{n=N}^{\infty} |a_n| |x|^{n-N} \leq K |x|^N$$

where $K = \sum_{n=N}^{\infty} |a_n| r^{n-N}$. □

The point here is that K, though dependent on N and r, does not depend on x. This gives us scope for applications, the value of N being chosen for the purpose in hand.

Example If $x \in \mathbf{R}$, $n \log(1 + x/n) \to x$ as $n \to \infty$ and $(1 + x/n)^n \to e^x$ as $n \to \infty$.

Solution: For $|y| < 1$, $\log(1 + y) = \sum_{n=1}^{\infty} (-1)^{n-1} y^n/n$ (Problem 6). Therefore, by Theorem 9.11, there is a constant K such that $|y| \leq \frac{1}{2} \Rightarrow$ $|\log(1 + y) - y| \leq K |y|^2$. Let $x \in \mathbf{R}$. Then if $n \geq 2|x|$, $|x/n| \leq \frac{1}{2}$ and

$|\log(1 + x/n) - x/n| \le K|x|^2/n^2$ whence $|n \log(1 + x/n) - x| \le K|x|^2/n$. It is now clear that $\forall n > \max(2|x|, K|x|^2/\varepsilon) \ |n \log(1 + x/n) - x| < \varepsilon$.

Since exp is continuous and $(1 + x/n)^n = \exp(n \log(1 + x/n))$ we see $(1 + x/n)^n \to e^x$ as $n \to \infty$. $\quad\square$

Problems

1. Calculate the radius of convergence of the following series, where there is one, or show the series converges for all $x \in \mathbf{R}$: $\sum x^n/n$, $\sum x^n/2^n$, $\sum((2n)!/(n!)^2)x^n$, $\sum n^n x^n$, $\sum((2n)!/(3n)!)x^n$, $\sum nx^{3n+1}$, $\sum 2^n x^{2n}$.

2. Show that the series $\sum x/(1 + |x|)^n$ is convergent for all real x, and find $\sum_{n=0}^{\infty} x/(1 + |x|)^n$. Notice that the sum is discontinuous at $x = 0$.

3. Let $\alpha > 0$. By using the result $x^{-1} \log x \to 0$ as $x \to \infty$, deduce that, for all sufficiently large x, $(\alpha \log x)/x < \frac{1}{2}$ and hence $\alpha \log x - x < -x/2$. Hence show that $x^\alpha e^{-x} \to 0$ as $x \to \infty$. (α here need not be an integer.) Show also that $x^{-\alpha} e^x \to \infty$ as $x \to \infty$.

4. Calculate $\lim_{x \to 0+} x^x$ and $\lim_{x \to \infty} x^{1/x}$.

5. Define $f:(-R, R) \to \mathbf{R}$ by $f(x) = \sum_{n=0}^{\infty} a_n x^n$ (where the series is presumed to converge). Show that $a_n = f^{(n)}(0)/n!$

6. Let $f(x) = \log(1 + x) \ (x > -1)$ and show that $f^{(n)}(0) = (-1)^{n-1}(n-1)!$ $(n \ge 1)$, $f(0) = 0$. This suggests we define $g:(-1, 1) \to \mathbf{R}$ by $g(x) = \sum_{n=1}^{\infty} (-1)^{n-1} x^n/n$. Check this series has radius of convergence 1 and that $g'(x) = 1/(1 + x)$. By considering $(1 + x)e^{-g(x)}$, or otherwise, show that $g(x) = \log(1 + x)$.

7. Define the two functions sin and cos by

$$\sin x = \sum_{n=0}^{\infty} (-1)^n x^{2n+1}/(2n + 1)!, \quad \cos x = \sum_{n=0}^{\infty} (-1)^n x^{2n}/(2n)!$$

(i) Show that both series converge for all real x and that $(d/dx)(\sin x) = \cos x$, $(d/dx)(\cos x) = -\sin x$.

(ii) Use the Cauchy Product Theorem to prove that $\forall x, y \in \mathbf{R}$, $\sin(x + y) = \sin x \cos y + \cos x \sin y$.

(iii) Consider y fixed, x variable in (ii) and deduce that $\cos(x + y) = \cos x \cos y - \sin x \sin y$.

(iv) Show that $\cos(0) = 1$, $\sin(0) = 0$, $\cos(-x) = \cos x$, $\sin(-x) = -\sin x$ and $(\sin x)^2 + (\cos x)^2 = 1$.

(v) In the proof of the Alternating Series Theorem in Chapter 5, we saw that if (a_n) is a decreasing sequence of positive numbers and $s_n = a_1 - a_2 + \cdots + (-1)^{n-1} a_n$ then $s_{2n} \le \sum_{j=1}^{\infty} (-1)^{j-1} a_j \le s_{2n+1}$. Adapt this, taking care with the indices, which start at 0, to show that $0 < x < 2 \Rightarrow (\sin x)/x > 0$. Deduce that $0 < x < 2 \Rightarrow \sin x > 0$.

(vi) Use (v) to show cos is strictly decreasing on $[0, 2]$ and, by a similar use of the Alternating Series Theorem, prove cos $2 < 0$. Deduce that there is exactly one real number ξ with $0 < \xi < 2$ for which $\cos \xi = 0$. *Define π by $\pi = 2\xi$, and calculate* $\sin(\pi/2)$, $\sin \pi$, $\cos \pi$.

(vii) Show that $\forall x \in \mathbf{R}$ $\sin(x + 2\pi) = \sin x$, $\cos(x + 2\pi) = \cos x$.

8. With $\sin x$ as defined above, show that $\lim_{x \to 0} (\sin x)/x = 1$.

9. Show that there is a constant K such that $|y| \leq \frac{1}{2} \Rightarrow$ $|\sqrt{(1 + y)} - (1 + y/2)| \leq K|y|^2$, and use this to show that $\lim_{x \to 0} (\sqrt{(1 + x)} - \sqrt{(1 - x)})/x = 1$.

10. Show that if $x > 0$, $0 \leq x - \log(1 + x) \leq x^2/2$, and deduce that if $a_n = (1 + \frac{1}{2} + \cdots + 1/n) - \log(n + 1)$ then (a_n) is increasing and bounded above. Deduce that $1 + \frac{1}{2} + \cdots + 1/n - \log n$ tends to a limit as $n \to \infty$.

11. Suppose that $y(x) = \sum_{n=0}^{\infty} a_n x^n$ $\forall x \in (-R, R)$ and that y satisfies the differential equation $x^2(dy/dx) = y - x$. Find the coefficients a_n and the radius of convergence of this series.

12* Define $f(x) = e^{-1/x}$ $(x > 0)$, $f(x) = 0$ $(x \leq 0)$. Show that f is differentiable at 0 and $f'(0) = 0$. Show also that for $x > 0$, $f'(x) = (1/x^2)e^{-1/x}$ and that f' is continuous at 0, and for each $n \in \mathbf{N}$ show that $f^{(n)}(x) = p_n(1/x)e^{-1/x}$ for some polynomial p_n of degree at most $2n$. Deduce that for all natural numbers n, f is n times differentiable at 0, and $f^{(n)}(0) = 0$.

CHAPTER 10

Integration

"But just as much as it is easy to find the differential of any given quantity, so it is difficult to find the integral of a given differential. Moreover, sometimes we cannot even say with certainty whether the integral of a given quantity can be found or not ...."

Johann Bernoulli (1691)

10.1 THE INTEGRAL

The idea of the integral as the area under a curve is an old one, and it was known before the rise of calculus. The great feature which established the calculus was the observation that integration and differentiation are related so that if it is possible to find a function F whose derivative is f then

$$\int_a^b f(x)\,dx = F(b) - F(a),$$

solving the problem of integrating f. This method of integration, by finding a function whose derivative is the function we wish to integrate, is sometimes called 'anti-differentiation'.

While this provides an easy method for dealing with many integrals, it is not effective in all cases. Since the process of finding a function whose derivative is f is fairly ad hoc, relying on noticing suitable substitutions and so on, it is of no help at all in those cases where we are unable to find the desired function. Indeed, it can be shown that certain functions, such as e^{-x^2}, do not have an indefinite integral which can be expressed in terms of algebraic functions, exponentials and logarithms. We need an idea of integral which will allow us to deal with all the functions likely to occur in mathematics, and one where, even if in specific cases we cannot evaluate the integral exactly, we can still calculate it approximately to within some prescribed accuracy. The method is to return to the idea of the area under a curve and work with that.

Let $f:[a,b] \to \mathbf{R}$ be a bounded function, and choose a finite sequence of numbers $x_0,\ldots,x_n$ such that $a = x_0 < x_1 < \cdots < x_n = b$. Then choose numbers M_i for which $x_{i-1} \le x \le x_i \Rightarrow f(x) \le M_i$. The intuitive idea of area now tells us that the area under the curve $y = f(x)$ between $x = x_{i-1}$ and $x = x_i$ will not exceed $M_i(x_i - x_{i-1})$, the area of the rectangle of height M_i and base $[x_{i-1}, x_i]$. Therefore the area under f between $x = a$ and $x = b$ will be no

114

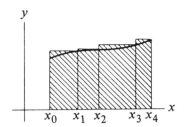

Fig. 10.1

more than $\sum_{i=1}^{n} M_i(x_i - x_{i-1})$. In this procedure we only required of M_i that it should satisfy $M_i \geq f(x)\; \forall x \in [x_{i-1}, x_i]$, which leaves us the option of choosing M_i very much larger than the values of $f(x)$. Obviously the most useful estimate of the area under f is obtained by choosing M_i as small as possible, that is, $M_i = \sup\{f(x) : x_{i-1} \leq x \leq x_i\}$. (Notice, in passing, that since f is given to be bounded, there is no doubt about the existence of the supremum.) The value of the sum $\sum_{i=1}^{n} M_i(x_i - x_{i-1})$ we obtain by this process will depend on the choice of the points $x_0, \ldots, x_n$ which we use to subdivide the interval $[a, b]$, and we would expect that, by taking more points of dissection, we can make the sum smaller, and thus a closer approximation to the area under the curve f. The independent variable here is the dissection of the interval, a complicated object, so we will not look for a limit but seek the 'smallest' value of these sums or, in case the set of all these sums has no smallest member, the infimum over all the sums obtained.

Before we rush on with this, notice that we could just as well approximate the integral from below. In this case, if $a = x_0 < x_1 < \cdots < x_n = b$ and $m_i = \inf\{f(x) : x_{i-1} \leq x \leq x_i\}$, then $\sum_{i=1}^{n} m_i(x_i - x_{i-1})$ should not exceed the area under the curve $y = f(x)$ between $x = a$ and $x = b$, and taking more points of dissection ought to make the sum closer to the area under the curve. This leads us to consider the supremum of the set of all sums $\sum_{i=1}^{n} m_i(x_i - x_{i-1})$.

These two ways of evaluating what we think of as the area under a curve are equally valid and either could be used as the definition of the integral, *provided they give the same answer*. Intuitively, it is hard to imagine the two processes giving different answers, at least for 'ordinary' functions, but we must not presume that there are no peculiar functions which would yield different answers. The definition below acknowledges this issue.

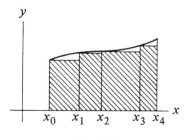

Fig. 10.2

Definition   Let $a < b$ and let $f:[a,b] \to \mathbf{R}$ be a bounded function. A **dissection** of the interval $[a,b]$ is a finite sequence of points $x_0,\ldots,x_n$ such that $a = x_0 < x_1 < \cdots < x_n = b$. Suppose that $D = \{x_0, x_1,\ldots,x_n\}$ is a dissection, then by the **upper sum**, $S(D)$, we mean

$$S(D) = \sum_{i=1}^{n} M_i(x_i - x_{i-1}),$$

where $M_i = \sup\{f(x): x_{i-1} \le x \le x_i\}$, and by the **lower sum**, $s(D)$, we mean

$$s(D) = \sum_{i=1}^{n} m_i(x_i - x_{i-1}),$$

where $m_i = \inf\{f(x): x_{i-1} \le x \le x_i\}$. Where it is necessary to distinguish the relevant function we shall denote these sums by $S(f;D)$ and $s(f;D)$ respectively.

We say that the function f is **integrable** (on $[a,b]$) if and only if $\inf\{S(D):D \text{ is a dissection}\} = \sup\{s(D):D \text{ is a dissection}\}$, and when f is integrable we denote the common value of these two quantities by $\int_a^b f$ or $\int_a^b f(x)\,dx$.

Lemma 10.1   Let $f:[a,b]\to\mathbf{R}$ be a bounded function and let D_1 and D_2 be two dissections of $[a,b]$. If D_2 contains every point of dissection that belongs to D_1 then $S(D_2) \le S(D_1)$ and $s(D_2) \ge s(D_1)$. Also

$$\sup\{s(D):D \text{ is a dissection}\} \le \inf\{S(D):D \text{ is a dissection}\}.$$

Proof: We shall prove that $S(D_2) \le S(D_1)$ in the case where D_2 contains one more point of dissection than D_1. The general case is obtained by applying this result a finite number of times, since D_2 can, in general, have only finitely many more points than D_1.

Let $D_1 = \{x_0, x_1,\ldots,x_n\}$ and $D_2 = \{x_0,\ldots,x_{j-1}, y, x_j,\ldots,x_n\}$ where $a = x_0 < x_1 < \cdots < x_{j-1} < y < x_j < \cdots < x_n = b$. Then $S(D_1) = \sum_{i=1}^{n} M_i(x_i - x_{i-1})$ where $M_i = \sup\{f(x):x_{i-1} \le x \le x_i\}$, $S(D_2)$ being given by a similar expression with the term $M_j(x_j - x_{j-1})$ replaced by $\bar{M}_j(y - x_{j-1}) + M_j'(x_j - y)$ where $\bar{M}_j = \sup\{f(x):x_{j-1} \le x \le y\}$ and $M_j' = \sup\{f(x):y \le x \le x_j\}$. Therefore

$$S(D_1) - S(D_2) = M_j(x_j - x_{j-1}) - \bar{M}_j(y - x_{j-1}) - M_j'(x_j - y)$$
$$= (M_j - \bar{M}_j)(y - x_{j-1}) + (M_j - M_j')(x_j - y) \ge 0$$

since $M_j \ge \bar{M}_j$ and $M_j \ge M_j'$ from their definitions.

The proof that $s(D_2) \ge s(D_1)$ is virtually identical.

Now let D_3 and D_4 be two typical dissections and form the dissection D_5 containing those points which are either in D_3 or in D_4. By what we have just proved, $S(D_5) \le S(D_3)$ and $s(D_5) \ge s(D_4)$. Moreover, from the definition of upper and lower sums, $s(D_5) \le S(D_5)$ (since $m_i \le M_i$ in the notation of the definition). Therefore $s(D_4) \le S(D_3)$.

Since D_3 and D_4 were typical in the above paragraph, we see that if we choose D_3, then for all dissections D_4, $s(D_4) \le S(D_3)$ hence $\sup\{s(D_4):D_4 \text{ is a dissection}\} \le S(D_3)$. Since this result is true for all D_3, $\sup\{s(D_4):D_4 \text{ is a dissection}\} \le \inf\{S(D_3):D_3 \text{ is a dissection}\}$. $\square$

Note: The inequality between two upper sums or between two lower sums requires that the two dissections be comparable, that is, one contains all the points of the other. The inequality between an upper sum and a lower sum does not require that the dissections be comparable; every upper sum is at least as large as every lower sum.

   This allows us a simple criterion for integrability:

Lemma 10.2   Let $f:[a,b] \to \mathbf{R}$ be a bounded function. f is integrable if and only if $\forall \varepsilon > 0 \; \exists$ a dissection D s.t. $S(D) - s(D) < \varepsilon$.

Proof: Let f be integrable and suppose that $\varepsilon > 0$. From the definitions of sup and inf, there exist dissections D_1 and D_2 satisfying $S(D_1) < \inf \{ S(D) : D$ is a dissection$\} + \varepsilon/2$ and $s(D_2) > \sup \{ s(D) : D$ is a dissection$\} - \varepsilon/2$. Let I denote the common value of $\inf \{ S(D) \}$ and $\sup \{ s(D) \}$ (f is integrable) and let D_3 be the dissection consisting of points which occur in D_1 or in D_2. Then $S(D_3) \le S(D_1) < I + \varepsilon/2 < s(D_2) + \varepsilon \le s(D_3) + \varepsilon$, yielding $S(D_3) - s(D_3) < \varepsilon$. Since $\varepsilon > 0$ was arbitrary we have proved one implication.

   Conversely, suppose that $\forall \varepsilon > 0 \; \exists D$ s.t. $S(D) - s(D) < \varepsilon$. Let $\varepsilon > 0$ and choose D_0 such that $S(D_0) - s(D_0) < \varepsilon$. Then

$$\inf \{ S(D) : D \text{ is a dissection} \} \le S(D_0) < s(D_0) + \varepsilon$$

$$\le \sup \{ s(D) : D \text{ is a dissection} \} + \varepsilon.$$

Since ε was arbitrary we deduce that $\forall \varepsilon > 0 \; \inf \{ S(D) \} < \sup \{ s(D) \} + \varepsilon$. Whence $\inf \{ S(D) \} \le \sup \{ s(D) \}$. The opposite inequality follows from Lemma 10.1, whence f is integrable. $\square$

Example   Let $f:[0,1] \to \mathbf{R}$ be defined by $f(x) = x$; f is integrable. Let $D_n = \{ 0, 1/n, 2/n, \ldots, 1 \}$, so, letting $x_j = j/n$, we have $M_j = \sup \{ f(x) : x_{j-1} \le x \le x_j \} = j/n$ and $m_j = \inf \{ f(x) : x_{j-1} \le x \le x_j \} = (j-1)/n$. Then $S(D_n) - s(D_n) = \sum_{j=1}^{n} (M_j - m_j)(x_j - x_{j-1}) = \sum_{j=1}^{n} 1/n^2 = 1/n$. Thus, given $\varepsilon > 0$, if we choose $n > 1/\varepsilon$ we obtain D_n for which $S(D_n) - s(D_n) < \varepsilon$.

   In fact, $\inf \{ S(D) :$ all dissections $D \} \le \inf \{ S(D_n) ; n \in \mathbf{N} \} = 1/2$ (since $S(D_n) = \sum_{j=1}^{n} j/n^2 = \frac{1}{2}(1 + 1/n))$. Also, since $s(D_n) = \frac{1}{2}(1 - 1/n)$, $\sup \{ s(D) :$ all dissections $D \} \ge \sup \{ s(D_n) : n \in \mathbf{N} \} = 1/2$, so

$$1/2 \le \sup \{ s(D) : \text{all } D \} \le \inf \{ S(D) : \text{all } D \} \le 1/2$$

and there is equality throughout, showing that $\int_0^1 f = 1/2$.

   The criterion of Lemma 10.2 in fact tells us more about approximations to the integral than we have so far stated.

Lemma 10.3   Let $f:[a,b] \to \mathbf{R}$ be integrable. Then if $\varepsilon > 0$ there is a dissection D_0 such that $S(D_0) - s(D_0) < \varepsilon$ and if D is any dissection whose points of subdivision include those of D_0, we have

$$\left| \int_a^b f - \sum_{j=1}^{n} f(\xi_j)(x_j - x_{j-1}) \right| < \varepsilon$$

where $D = \{ x_0, \ldots, x_n \}$ and the ξ_j are numbers satisfying $x_{j-1} \le \xi_j \le x_j$ $(j = 1, 2, \ldots, n)$.

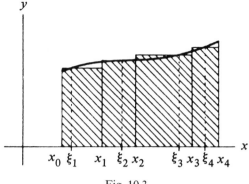

Fig. 10.3

Proof: Let $\varepsilon > 0$. By Lemma 10.2 there is dissection D_0 with
$S(D_0) - s(D_0) < \varepsilon$. Choose D to be a dissection containing all the points of
D_0, so that $S(D) \leq S(D_0) < s(D_0) + \varepsilon \leq \int_a^b f + \varepsilon$ and, similarly, $s(D) > \int_a^b f - \varepsilon$.
Now with $x_{j-1} \leq \xi_j \leq x_j$ we have
$m_j = \inf\{f(x) : x_{j-1} \leq x \leq x_j\} \leq f(\xi_j) \leq \sup\{f(x) : x_{j-1} \leq x \leq x_j\} = M_j$ so
$s(D) \leq \sum_{j=1}^n f(\xi_j)(x_j - x_{j-1}) \leq S(D)$, whence

$$\int_a^b f - \varepsilon < \sum_{j=1}^n f(\xi_j)(x_j - x_{j-1}) < \int_a^b f + \varepsilon,$$

which is what we require.   □

The definition of integral that we have made has the virtue that it is very
general, but as a technique for calculating the value of the integral exactly,
it is far from good. Our programme will be to show that the class of functions
which have integrals is usefully wide, then to establish the properties of the
integral, with the hope that the properties will allow us a reasonable method
of calculating most integrals that arise.

Lemma 10.4   If $f : [a, b] \to \mathbf{R}$ is either increasing or decreasing, it is integrable.

Proof: We shall prove the case where f is increasing; the other is virtually
identical.

Suppose that $D = \{x_0, \dots, x_n\}$ is a dissection of $[a, b]$. Then, because f is
increasing, $\sup\{f(x) : x_{j-1} \leq x \leq x_j\} = f(x_j)$ and
$\inf\{f(x) : x_{j-1} \leq x \leq x_j\} = f(x_{j-1})$ so $S(D) - s(D) =$
$\sum_{j=1}^n (f(x_j) - f(x_{j-1}))(x_j - x_{j-1})$; the issue is to show that, given $\varepsilon > 0$, a
dissection can be chosen for which this quantity is less than ε.

Let $\varepsilon > 0$. Assuming $f(b) \neq f(a)$, so $f(b) > f(a)$ since f is increasing, choose
the dissection $D = \{x_0, \dots, x_n\}$ such that, for each j, $x_j - x_{j-1} < \varepsilon/(f(b) - f(a))$,
which is obviously possible if $[a, b]$ is divided into sufficiently many equal
parts, for example. Then

$$S(D) - s(D) = \sum_{j=1}^n (f(x_j) - f(x_{j-1}))(x_j - x_{j-1})$$

$$< \{\varepsilon/(f(b) - f(a))\} \sum_{j=1}^n (f(x_j) - f(x_{j-1})) = \varepsilon.$$

Since $\varepsilon > 0$ was arbitrary, Lemma 10.2 now shows that f is integrable. If $f(b) = f(a)$, then f is constant, because increasing, so f is integrable. ☐

In the proof just given, $S(D) - s(D)$ was expressed in the form $\sum_{j=1}^{n}(M_j - m_j)(x_j - x_{j-1})$ and we proved that this was small by making all of the $(x_j - x_{j-1})$ small and ensuring that we knew $\sum(M_j - m_j)$. The alternative tactic of ensuring that, for all j, $M_j - m_j$ is small seems useful since we know $\sum(x_j - x_{j-1})$. This requires a little more work. Now

$$M_j - m_j = \sup\{f(x): x \in [x_{j-1}, x_j]\} - \inf\{f(y): y \in [x_{j-1}, x_j]\}$$
$$= \sup\{f(x): x \in [x_{j-1}, x_j]\} + \sup\{-f(y): y \in [x_{j-1}, x_j]\}$$
$$= \sup\{f(x) - f(y): x, y \in [x_{j-1}, x_j]\}$$
$$= \sup\{|f(x) - f(y)|: x, y \in [x_{j-1}, x_j]\}.$$

We therefore define the **oscillation** of f on the set A, written $\underset{A}{\operatorname{osc}} f$, by

$$\underset{A}{\operatorname{osc}} f = \sup\{|f(x) - f(y)|: x, y \in A\}$$
$$= \sup\{f(x): x \in A\} - \inf\{f(y): y \in A\}.$$

If f is continuous at c and $\varepsilon > 0$ then we know that there is a $\delta > 0$ for which $x \in (c - \delta, c + \delta) \Rightarrow |f(x) - f(c)| < \varepsilon/2$ whence $x, y \in (c - \delta, c + \delta) \Rightarrow$ $|f(x) - f(y)| < \varepsilon$, so that $\underset{(c-\delta,c+\delta)}{\operatorname{osc}} f \leq \varepsilon$. More importantly, we may extend this result:

Theorem 10.5   Let $a < b$ and $f:[a, b] \to \mathbf{R}$ be continuous. Then if $\varepsilon > 0$ there is a dissection $D = \{x_0, \ldots, x_n\}$ of $[a, b]$ such that, for each j, $\underset{[x_{j-1}, x_j]}{\operatorname{osc}} f \leq \varepsilon$.

Proof: Let $\varepsilon > 0$, By our remarks above, $\exists \delta > 0$ s.t. $\underset{[a, a+\delta)}{\operatorname{osc}} f \leq \varepsilon$ so if we choose $a < x < a + \delta$, the trivial dissection of $[a, x]$ has the property that the oscillation on its one subinterval does not exceed ε.

Let $A = \{x \in [a, b]: [a, x]$ has a dissection D such that on every subinterval of D, $\operatorname{osc} f \leq \varepsilon\}$. The previous paragraph shows that for some $\delta > 0$, $[a, a + \delta) \subset A$ so that A is non-empty and $\sup A > a$. ($\sup A$ exists since A is also bounded above by b.) Let $\xi = \sup A$.

Firstly we show that $\xi = b$. Suppose that $\xi < b$. Then since f is continuous at ξ, $\exists \delta > 0$ s.t. $\underset{(\xi-\delta,\xi+\delta)}{\operatorname{osc}} f \leq \varepsilon$. Since $\xi - \delta < \xi$ there is an element, say x, of A with $x > \xi - \delta$. Then there is a dissection, say $\{x_0, \ldots, x_n\}$, of $[a, x]$ with $\underset{[x_{j-1}, x_j]}{\operatorname{osc}} f \leq \varepsilon$ for each j. Let $x_{n+1} = \xi + \delta/2$, so that $\underset{[x_n, x_{n+1}]}{\operatorname{osc}} f \leq \varepsilon$ since $[x_n, x_{n+1}] \subset (\xi - \delta, \xi + \delta)$ and every subinterval of the dissection $\{x_0, \ldots, x_{n+1}\}$ has $\operatorname{osc} f \leq \varepsilon$, so $x_{n+1} = \xi + \delta/2 \in A$. But this is a contradiction since $\xi + \delta/2 > \sup A$.

We know $\xi \geq b$, hence $\xi = b$ since b is an upper bound for A. We need to show $b \in A$. Because f is continuous at b, $\exists \delta > 0$ s.t. $\underset{(b-\delta,b]}{\operatorname{osc}} f \leq \varepsilon$. Since $b - \delta < b = \sup A$, there is a member of A, say x, with $x > b - \delta$. Then there

is a dissection $\{x_0,\ldots,x_n\}$ of $[a,x]$ for which, for $j=1,\ldots,n,$ $\operatorname*{osc}_{[x_{j-1},x_j]} f \leq \varepsilon$
whence $\operatorname{osc} f \leq \varepsilon$ on each subinterval of $\{x_0,\ldots,x_n,b\}$, and $b\in A$, as required.

Since $\varepsilon>0$ was arbitrary we have proved the result for all $\varepsilon>0$. $\square$

Corollary If $f:[a,b]\to\mathbf{R}$ is continuous, it is integrable.

Proof: Let $\varepsilon>0$. By Theorem 10.5 there is a dissection $D=\{x_0,\ldots,x_n\}$ of $[a,b]$ with $\operatorname*{osc}_{[x_{j-1},x_j]} f \leq \varepsilon/(2(b-a))$. Then

$$S(D)-s(D)=\sum_{j=1}^{n}(M_j-m_j)(x_j-x_{j-1})=\sum_{j=1}^{n}\operatorname*{osc}_{[x_{j-1},x_j]} f\cdot(x_j-x_{j-1})$$

$$\leq \varepsilon/(2(b-a))\cdot\sum_{j=1}^{n}(x_j-x_{j-1})=\varepsilon/2<\varepsilon.$$

By Lemma 10.2, f is integrable. $\square$

Although it is not about integration, it is useful to state another important corollary of Theorem 10.5 here.

Theorem 10.6 (The Uniform Continuity Theorem) Let $f:[a,b]\to\mathbf{R}$ be continuous. Then for all $\varepsilon>0$ there is a $\delta>0$ such that

$$\forall x,y\in[a,b] \quad |x-y|<\delta\Rightarrow|f(x)-f(y)|<\varepsilon.$$

Proof: Suppose that $\varepsilon>0$. Then, since $\varepsilon/3>0$, Theorem 10.5 ensures that there is a dissection $D=\{x_0,\ldots,x_n\}$ such that, for $j=1,\ldots,n,$ $\operatorname*{osc}_{[x_{j-1},x_j]} f\leq\varepsilon/3$. Let $\delta=\min(x_1-x_0,\ldots,x_n-x_{n-1})$, so $\delta>0$, and if $|x-y|<\delta$ then x and y either lie in the same subinterval of D or lie in adjacent subintervals. Let c be the common endpoint of the two adjacent intervals to which x and y belong, or an endpoint if x and y belong to the same interval; then

$$|f(x)-f(y)|\leq|f(x)-f(c)|+|f(c)-f(y)|\leq\varepsilon/3+\varepsilon/3<\varepsilon$$

as x and c belong to the same subinterval and so do c and y. We have shown that $\forall\varepsilon>0\ \exists\delta>0$ s.t. $\forall x,y\in[a,b]\ |x-y|<\delta\Rightarrow|f(x)-f(y)|<\varepsilon.$ $\square$

Remarks: Notice what the Uniform Continuity Theorem says: if f is defined and continuous on a closed bounded interval then, given $\varepsilon>0$, there is a $\delta>0$ (*and the same δ at all points*) such that for any two points of $[a,b]$ less than δ apart, $|f(x)-f(y)|<\varepsilon$. This is a stronger statement than continuity, since if f is continuous we know that f is continuous at each point y, that is, given $\varepsilon>0$ and $y\in[a,b]$, there is a $\delta>0$ such that if $|x-y|<\delta$ then $|f(x)-f(y)|<\varepsilon$; in this case the existence of δ is guaranteed once we fix ε and y, so δ may depend on y.

Theorems 10.5 and 10.6 depend on the fact that $[a,b]$ is closed and they are false if this is relaxed. Consider $f(x)=1/x$ where $f:(0,1]\to\mathbf{R}$. Let $\delta>0$ and choose $n\in\mathbf{N}$ with $1/n<\delta$. Then if we set $x=1/(2n)$ and $y=1/n$ we have $|x-y|=1/(2n)<\delta$ and $|f(x)-f(y)|=n\geq1$. Thus $\forall\delta>0\ \exists x,y\in(0,1)$ with $|x-y|<\delta$ and $|f(x)-f(y)|\geq1$ so f does not have the 'uniform' continuity property above (since it does not hold for $\varepsilon=1$).

Theorem 10.7 Let f and g be integrable functions on $[a, b]$ and λ be a real number. Then (i) $f + g$, $\lambda f, fg$ and $|f|$ are all integrable, (ii) if there is a constant $m > 0$ for which $\forall x \in [a, b]$ $|g(x)| \geq m$ then $1/g$ is integrable, and (iii) $\int_a^b (f + g) = \int_a^b f + \int_a^b g$ and $\int_a^b \lambda f = \lambda \int_a^b f$.

Proof: To prove the various functions integrable, we shall use Lemma 10.2. To do this we shall need to show, for a given function h and $\varepsilon > 0$, that for some D, $S(h; D) - s(h; D) = \sum_{j=1}^n (\underset{[x_{j-1}, x_j]}{\operatorname{osc}} h)(x_j - x_{j-1}) < \varepsilon$. We shall do this by relating $\operatorname{osc} h$ to the oscillation of the functions defining it.

$$\underset{[x_{j-1}, x_j]}{\operatorname{osc}} (f + g) = \sup \{|f(x) + g(x) - (f(y) + g(y))| : x, y \in [x_{j-1}, x_j]\}$$

$$\leq \sup \{|f(x) - f(y)| + |g(x) - g(y)| : x, y \in [x_{j-1}, x_j]\}$$

$$\leq \sup \{|f(x) - f(y)| : x, y \in [x_{j-1}, x_j]\}$$

$$+ \sup \{|g(x) - g(y)| : x, y \in [x_{j-1}, x_j]\}$$

$$= \underset{[x_{j-1}, x_j]}{\operatorname{osc}} f + \underset{[x_{j-1}, x_j]}{\operatorname{osc}} g .$$

Let $\varepsilon > 0$. Since f and g are integrable on $[a, b]$ there are dissections D_1 and D_2 such that $S(f; D_1) - s(f; D_1) < \varepsilon/2$ and $S(g; D_2) - s(g; D_2) < \varepsilon/2$. Letting D_3 be the dissection whose points of subdivision are those occurring in either D_1 or D_2, we have $S(f; D_3) - s(f; D_3) \leq S(f; D_1) - s(f; D_1) < \varepsilon/2$ and, similarly, $S(g; D_3) - s(g; D_3) < \varepsilon/2$. Thus if $D_3 = \{x_0, \ldots, x_n\}$,

$$S(f + g; D_3) - s(f + g; D_3) = \sum_{j=1}^n \underset{[x_{j-1}, x_j]}{\operatorname{osc}} (f + g) \cdot (x_j - x_{j-1})$$

$$\leq \sum_{j=1}^n (\underset{[x_{j-1}, x_j]}{\operatorname{osc}} f + \underset{[x_{j-1}, x_j]}{\operatorname{osc}} g)(x_j - x_{j-1})$$

$$= S(f; D_3) - s(f; D_3) + S(g; D_3) - s(g; D_3)$$

$$< \varepsilon.$$

Since $\varepsilon > 0$ was arbitrary we have shown $f + g$ is integrable.

For λf, we notice that $\operatorname{osc} \lambda f = |\lambda| \operatorname{osc} f$. Then, for $\varepsilon > 0$, choose D_1 such that $S(f; D_1) - s(f; D_1) < \varepsilon/|\lambda|$, whence $S(\lambda f; D_1) - s(\lambda f; D_1) < \varepsilon$. Thus λf is integrable, since the above holds for all $\varepsilon > 0$. (The case $\lambda = 0$ is trivial.)

Let M_f and M_g be upper bounds for $|f|$ and $|g|$ respectively. Then for all

$$x, y \in [a, b], \quad |f(x)g(x) - f(y)g(y)| \leq |f(x)||g(x) - g(y)| + |f(x) - f(y)||g(y)|$$

$$\leq M_f|g(x) - g(y)| + M_g|f(x) - f(y)|.$$

Therefore, on all subsets A of $[a, b]$, $\underset{A}{\operatorname{osc}} fg \leq M_f \underset{A}{\operatorname{osc}} g + M_g \underset{A}{\operatorname{osc}} f$, giving $S(fg; D) - s(fg; D) \leq M_f (S(g; D) - s(g; D)) + M_g (S(f; D) - s(f; D))$. Given $\varepsilon > 0$ it is only necessary to choose D_1 and D_2 such that $S(f; D_1) - s(f; D_1) < \varepsilon/(M_f + M_g)$ and $S(g; D_2) - s(g; D_2) < \varepsilon/(M_f + M_g)$, giving $S(fg; D_3) - s(fg; D_3) < \varepsilon$ where D_3 contains all the points of D_1 and D_2.

Since $||f(x)| - |f(y)|| \leq |f(x) - f(y)|$, $\operatorname{osc} |f| \leq \operatorname{osc} f$, and the usual routine shows that $|f|$ is integrable.

The integrability of $1/g$ follows from noticing that, with the given condition, $|1/g(x) - 1/g(y)| \leq |g(x) - g(y)|/m^2$, so $\operatorname{osc}(1/g) \leq (\operatorname{osc} g)/m^2$.

For part (iii), we need to estimate the values of the integrals. Let $\varepsilon > 0$. Then choose D_1 and D_2 so that $S(f; D_1) - s(f; D_1) < \varepsilon/2$ and $S(g; D_2) - s(g; D_2) < \varepsilon/2$. Let D_3 be the dissection whose points of subdivision are those occurring in either D_1 or D_2. Then

$$\int_a^b (f + g) \le S(f + g; D_3) \le S(f; D_3) + S(g; D_3) < s(f; D_3) + s(g; D_3) + \varepsilon$$

$$\le \int_a^b f + \int_a^b g + \varepsilon.$$

Thus

$$\forall \varepsilon > 0 \int_a^b (f + g) \le \int_a^b f + \int_a^b g + \varepsilon, \quad \text{whence} \quad \int_a^b (f + g) \le \int_a^b f + \int_a^b g.$$

Arguing similarly in terms of $s(f + g; D_3)$ we may show the reverse inequality also holds and so $\int_a^b (f + g) = \int_a^b f + \int_a^b g$.

To evaluate $\int_a^b \lambda f$, notice that $S(\lambda f; D) = \lambda S(f; D)$ if $\lambda \ge 0$ and $S(\lambda f; D) = \lambda s(f; D)$ if $\lambda \le 0$, from which the equality of $\int_a^b \lambda f$ and $\lambda \int_a^b f$ may be seen from the definition of integral. □

This gives us a wide class of integrable functions: those which are increasing or continuous or which can be expressed in terms of sums, products, etc. of such functions. To find a function which is not integrable we have to resort to something fairly esoteric.

Example Define $f:[0, 1] \to \mathbf{R}$ by $f(x) = 1$ if x is rational and $f(x) = 0$ if x is irrational. f is not integrable. To see this, let $D = \{x_0, x_1, \ldots, x_n\}$. Then since each interval $[x_{j-1}, x_j]$ contains both rational and irrational numbers (as $x_{j-1} < x_j$), $\sup\{f(x):x \in [x_{j-1}, x_j]\} = 1$ and $\inf\{f(x):x \in [x_{j-1}, x_j]\} = 0$. This gives $S(f; D) = \sum_{j=1}^n (x_j - x_{j-1}) = 1$ and $s(f; D) = 0$. Thus $\inf S(f; D) = 1 \ne 0 = \sup s(f; D)$.

Despite the existence of non-integrable functions, we shall not find this presents a problem. The issue only becomes troublesome when one wishes to consider limits of sequences of functions, which we shall not do in this book.

Lemma 10.8 Let $a < b < c$. A function $f:[a, c] \to \mathbf{R}$ is integrable on $[a, c]$ if and only if it is integrable on both $[a, b]$ and $[b, c]$. When f is integrable on $[a, c]$, $\int_a^c f = \int_a^b f + \int_b^c f$.

Proof: Suppose that f is integrable on $[a, b]$ and $[b, c]$. Let $\varepsilon > 0$. Choose dissections $D_1 = \{x_0, \ldots, x_n\}$ of $[a, b]$ and $D_2 = \{y_0, \ldots, y_m\}$ of $[b, c]$ satisfying $S(D_1) - s(D_1) < \varepsilon/2$ and $S(D_2) - s(D_2) < \varepsilon/2$. Then $D_3 = \{x_0, \ldots, x_n, y_1, \ldots, y_m\}$ is a dissection of $[a, c]$ and $S(D_3) = S(D_1) + S(D_2)$, $s(D_3) = s(D_1) + s(D_2)$, whence $S(D_3) - s(D_3) < \varepsilon$. Thus

$$\int_a^c f \le S(D_3) = S(D_1) + S(D_2) < s(D_1) + s(D_2) + \varepsilon \le \int_a^b f + \int_b^c f + \varepsilon,$$

and

$$\int_a^c f \geq s(D_3) = s(D_1) + s(D_2) > S(D_1) + S(D_2) - \varepsilon \geq \int_a^b f + \int_b^c f - \varepsilon.$$

Since ε was arbitrary, we see that f is integrable on $[a, b]$ and $\int_a^c f = \int_a^b f + \int_b^c f$ (the equation because both inequalities between the left and right-hand sides hold).

Conversely, suppose that f is integrable on $[a, c]$ and let $\varepsilon > 0$. Then there is a dissection D for which $S(D) - s(D) < \varepsilon$. Let D_3 be the dissection obtained from D by inserting b as a point of dissection, so $S(D_3) - s(D_3) < \varepsilon$. Now let D_1 consist of those points of dissection of D_3 lying in $[a, b]$, and D_2 consist of those in $[b, c]$. Since $(S(D_1) - s(D_1)) + (S(D_2) - s(D_2)) = S(D_3) - s(D_3) < \varepsilon$ and both brackets on the left are non-negative, $S(D_1) - s(D_1) < \varepsilon$ and $S(D_2) - s(D_2) < \varepsilon$. Since ε was arbitrary, f is integrable on $[a, b]$ and $[b, c]$. The equality of the integrals follows from the first part. $\square$

Lemma 10.9 Suppose that $f(x) = g(x)$ for all but finitely many points of $[a, b]$. Then f is integrable if and only if g is, and $\int_a^b f = \int_a^b g$ when either function is integrable. In particular, if $h(x) = 0$ at all but finitely many points of $[a, b]$ then $\int_a^b h = 0$.

Proof: We shall prove a special case and deduce the general one from it. Suppose $h(x) = 0$ for all $x \in [c, d]$ except one of the endpoints; for definiteness, let $h(c) = \alpha > 0$. Then if $D_\delta = \{c, c + \delta, d\}$, $S(D_\delta) = \alpha \delta$ and $s(D_\delta) = 0$, so

$$\inf \{S(D) : D \text{ any dissection}\} \leq \inf \{S(D_\delta) : \delta > 0\} = 0 = \sup s(D).$$

Thus h is integrable and $\int_c^d h = 0$. The result for a function h which is non-zero at finitely many points may now be deduced by splitting the interval $[a, b]$ at the points at which h is non-zero, observing that the integral of h on each subinterval is zero and using Lemma 10.8.

If $f(x) = g(x)$ for all but finitely many points x, then $f - g$ is zero at all but finitely many points, so the preceding paragraph shows that $f - g$ is integrable and $\int_a^b (f - g) = 0$. The rest is easy. $\square$

Lemma 10.10 Suppose that f and g are integrable on $[a, b]$ and that $\forall x \in [a, b]$ $f(x) \geq g(x)$; then $\int_a^b f \geq \int_a^b g$. If h is integrable on $[a, b]$ then $|\int_a^b h| \leq \int_a^b |h|$.

Proof: If f and g are integrable and $\forall x \in [a, b]$ $f(x) \geq g(x)$, then $f - g$ is integrable and $\forall x \in [a, b]$ $f(x) - g(x) \geq 0$. Thus, on each subinterval of a dissection $\inf (f - g) \geq 0$, so, for all dissections D, $s(f - g; D) \geq 0$. Thus $\int_a^b (f - g) \geq s(f - g; D) \geq 0$.

$$\int_a^b f = \int_a^b (f - g) + \int_a^b g \geq \int_a^b g.$$

For the last part, if h is integrable, so is $|h|$ and $\forall x \in [a, b]$

$-|h(x)| \leq h(x) \leq |h(x)|$ so $-\int_a^b |h| \leq \int_a^b h \leq \int_a^b |h|$ from which we see that $|\int_a^b h| \leq \int_a^b |h|$. $\square$

Having finished the technicalities, we may proceed to the substantial results. To simplify matters, we shall **define** $\int_a^b f = -\int_b^a f$ where $b < a$. This does not affect the truth of the results already stated except in the case of Lemma 10.10 where the inequalities presume $a < b$. Lemma 10.10 could be re-stated, for example, as $|\int_a^b h| \leq \int_\alpha^\beta |h|$ where $\alpha = \min(a, b)$ and $\beta = \max(a, b)$. We also define $\int_a^a f = 0$ for all a.

Theorem 10.11 (The Mean Value Theorem for Integrals) Suppose that $f:[a,b] \to \mathbf{R}$ is continuous. There is a point $\xi \in [a, b]$ for which

$$\int_a^b f = (b - a)f(\xi).$$

Proof: Let $m = \inf\{f(x):a \leq x \leq b\}$ and $M = \sup\{f(x):a \leq x \leq b\}$, so, by the inequality result in Lemma 10.10, $m(b - a) \leq \int_a^b f \leq M(b - a)$. Therefore $(b - a)^{-1}\int_a^b f$ is intermediate between m and M. Since f is continuous, m and M are values attained by f in $[a, b]$, hence by the Intermediate Value Theorem $\exists \xi \in [a, b]$ with $f(\xi) = (b - a)^{-1}\int_a^b f$. (Thus assumes $a < b$; if not, consider $\int_b^a f$.) $\square$

Theorem 10.12 (The Fundamental Theorem of Calculus) Let $f:[a, b] \to \mathbf{R}$ be continuous and define F by $F(x) = \int_a^x f(t)\,dt$. Then F is differentiable and $F' = f$. Conversely, suppose that g is continuous on $[a, b]$ and differentiable on (a, b). Then if g' is integrable, $\int_a^b g'(t)\,dt = g(b) - g(a)$.

Proof: Choose $x \in [a, b]$ and fix it. Then for all h for which $x + h \in [a, b]$

$$F(x + h) - F(x) = \int_x^{x+h} f(t)\,dt.$$

By the Mean Value Theorem there is a ξ between x and $x + h$ for which $\int_x^{x+h} f(t)\,dt = hf(\xi)$.

Choose $\varepsilon > 0$. By the continuity of f at x, $\exists \delta > 0$ s.t. $|y - x| < \delta \Rightarrow |f(y) - f(x)| < \varepsilon$. For this δ, if $0 < |h| < \delta$ then the ξ above satisfies $|\xi - x| \leq |h| < \delta$, so

$$\left| \frac{F(x + h) - F(x)}{h} - f(x) \right| = |f(\xi) - f(x)| < \varepsilon.$$

Since ε was arbitrary, we have shown $F'(x) = f(x)$ (the derivative being the one-sided limit if $x = a$ or b). Since x was typical we have finished the first part.

For the second part, let g be as stated and $\varepsilon > 0$. By the integrability of g', there is a dissection $D = \{c_0, c_1, \ldots, c_n\}$ for which $S(g'; D) - s(g'; D) < \varepsilon$. By the Mean Value Theorem, for each j there is a $\xi_j \in [c_{j-1}, c_j]$ with $g(c_j) - g(c_{j-1}) = g'(\xi_j)(c_j - c_{j-1})$, so $g(b) - g(a) = \sum_{j=1}^n g'(\xi_j)(c_j - c_{j-1})$. This last expression lies between $S(g'; D)$ and $s(g'; D)$ so

$$\int_a^b g' - \varepsilon \leq S(g'; D) - \varepsilon < s(g'; D) \leq g(b) - g(a)$$

$$\leq S(g'; D) < s(g'; D) + \varepsilon \leq \int_a^b g' + \varepsilon$$

whence $g(b) - g(a)$ and $\int_a^b g'$ differ by less than ε. Since ε was arbitrary, $\int_a^b g' = g(b) - g(a)$.   □

Notice that the second part above has the (essential) requirement that the derivative be integrable, and that it relates integration to the 'calculus' method of finding a function whose derivative is the function to be integrated. Thus if the derivative is sufficiently well behaved, the integration 'undoes' the action of differentiating. The first part is slightly different; if f is continuous, it guarantees F is differentiable and differentiation 'undoes' the action of integrating.

Corollary 1 (Differentiation under the Integral sign) Let $\phi(x, t) = \sum_{j=1}^n f_j(x) g_j(t)$ where, for each j, f_j is differentiable and g_j is continuous. Then

$$\frac{d}{dx} \left(\int_a^x \phi(x, t) \, dt \right) = \phi(x, x) + \int_a^x \frac{\partial \phi}{\partial x}(x, t) \, dt.$$

Proof: By the Theorem and the rules for differentiating a product,

$$\frac{d}{dx} \left\{ \int_a^x f_j(x) g_j(t) \, dt \right\} = \frac{d}{dx} \left\{ f_j(x) \int_a^x g_j(t) \, dt \right\} = f_j(x) g_j(x) + f_j'(x) \int_a^x g_j(t) \, dt$$

$$= f_j(x) g_j(x) + \int_a^x f_j'(x) g_j(t) \, dt.$$

The result follows on summation.   □

Corollary 2   Let $h > 0$ and $g: [a - h, a + h] \to \mathbf{R}$ be continuous and define G_n by

$$G_n(x) = \frac{1}{(n-1)!} \int_a^x (x - t)^{n-1} g(t) \, dt \quad (n = 1, 2, \ldots).$$

Then $G_n(a) = G_n'(a) = \cdots = G_n^{(n-1)}(a) = 0$ and $G_n^{(n)} = g$.

Proof: We prove the result by induction. The Fundamental Theorem gives the desired properties of G_1.

Suppose the result is true for $n \in \mathbf{N}$. Then, since $(x - t)^n$ may be written as a sum of functions of the form $x^j t^{n-j}$, we may apply differentiation under the integral sign to G_{n+1} to obtain

$$G_{n+1}'(x) = \frac{1}{n!}(x - x)^n g(x) + \frac{1}{n!} \int_a^x n(x - t)^{n-1} g(t) \, dt = G_n(x),$$

and $G_{n+1}(a) = 0$. By the assumed result about G_n, $G_{n+1}^{(n+1)} = G_n^{(n)} = g$ and $G_{n+1}^{(j)}(a) = 0$ ($j = 1, \ldots, n$), so the result also holds for $n + 1$. $\square$

This Corollary gives a useful, but perhaps unexpected, method for calculating the result of n indefinite integration operations by performing a single integral. This yields another version of Taylor's Theorem where we obtain an exact formula for the remainder.

Corollary 3   Suppose that $f:[a - h, a + h] \to \mathbf{R}$ has derivatives of order up to and including n and that $f^{(n)}$ is continuous. Then if $a - h \le x \le a + h$,

$$f(x) = f(a) + (x - a)f'(a) + \cdots + (x - a)^{n-1}f^{(n-1)}(a)/(n - 1)! + R_n(x)$$

where

$$R_n(x) = \int_a^x \frac{(x - t)^{n-1}}{(n - 1)!} f^{(n)}(t)\,\mathrm{d}t.$$

Proof: Define the function $R_n(x)$ by the formula given so, by Corollary 2, $R_n^{(n)} = f^{(n)}$ whence $R_n - f$ is a polynomial of degree at most $n - 1$. Let $f(x) - R_n(x) = \sum_{j=0}^{n-1} \alpha_j(x - a)^j$. Then, for $j = 0, \ldots, n - 1$

$$f^{(j)}(a) - R_n^{(j)}(a) = f^{(j)}(a) = \alpha_j \cdot j!$$

Substituting these values of α_j gives the required result. $\square$

We shall postpone, for the moment, applications of these results and finish the task of relating the new concept of integration to the old, that is, show that all of the techniques used in the anti-differentiation view of integral apply to our integral, so we have lost nothing, and gained much, in our new attack. The last two matters to be settled in this direction are integration by parts and by substitution.

Lemma 10.13 (Integration by Parts)   Suppose f and g both have a continuous derivative. Then $\int_a^b fg' = \{f(b)g(b) - f(a)g(a)\} - \int_a^b f'g$.

Proof: $(fg)' = f'g + fg'$, so since $(fg)'$ is continuous, $\int_a^b (fg)' = f(b)g(b) - f(a)g(a)$. $\square$

The customary use of integration by parts to find $\int_a^b f\phi$ would use $g = \Phi$, where $\Phi' = \phi$.

Lemma 10.14 (Integration by Substitution)   Let f, ϕ and ϕ' be continuous where $f:[a, b] \to \mathbf{R}$, $\phi:[\alpha, \beta] \to [a, b]$. Then if ϕ possesses an inverse function ϕ^{-1},

$$\int_a^b f(x)\,\mathrm{d}x = \int_{\phi^{-1}(a)}^{\phi^{-1}(b)} f(\phi(t))\phi'(t)\,\mathrm{d}t.$$

Even if ϕ does not have an inverse function,

$$\int_{\phi(\alpha)}^{\phi(\beta)} f(x)\,\mathrm{d}x = \int_{\alpha}^{\beta} f(\phi(t))\phi'(t)\,\mathrm{d}t.$$

[Notice that these are exactly the results obtained from calculus by substituting $x = \phi(t)$, $dx = \phi'(t)\, dt$.]

Proof: We prove the second of these, since the first is a special case of it. Define

$$F(y) = \int_{\phi(\alpha)}^{\phi(y)} f(x)\, dx \quad \text{and} \quad G(y) = \int_{\alpha}^{y} f(\phi(t))\phi'(t)\, dt.$$

Then $F(y) = J(\phi(y))$ where $J(y) = \int_{\phi(\alpha)}^{y} f(x)\, dx$, so, using the Fundamental Theorem, which applies since all the integrands are continuous, we have, for all y, $F'(y) = J'(\phi(y))\phi'(y) = f(\phi(y))\phi'(y) = G'(y)$. Since $F' - G' = 0$, for all y, $F(y) - G(y) = F(\alpha) - G(\alpha) = 0$. In particular, $F(\beta) = G(\beta)$, the required result. $\square$

10.2 APPROXIMATING THE VALUE OF AN INTEGRAL

In this section we shall assume that f possesses as many derivatives as are needed for the calculation in hand and that these derivatives are continuous.

Suppose that $h > 0$ and that we wish to estimate the value of $\int_{a}^{a+h} f$. An obvious first estimate is $h \cdot (f(a) + f(a + h))/2$, the area of the shaded region in Fig. 10.4, but for this to be of much use we need to know how accurate an estimate it is. For this we proceed as follows:

Let $F(x) = \int_{a}^{x} f$, so $F' = f$, $F'' = f'$, etc., and we have $F(a + h) = F(a) + hf(a) + \frac{1}{2}h^2 f'(a) + R_1$, and $f(a + h) = f(a) + hf'(a) + R_2$, where R_1 and R_2 can be evaluated using Taylor's Theorem. Then

$$\int_{a}^{a+h} f = F(a + h) - F(a) = hf(a) + \tfrac{1}{2}h^2 f'(a) + R_1$$

$$= \tfrac{1}{2}h(f(a) + f(a + h)) + R_1 - \tfrac{1}{2}hR_2,$$

so the error in the estimate is $R_1 - \frac{1}{2}hR_2$.

At this point the ordinary form of remainder in Taylor's Theorem springs to mind, giving $R_1 = h^3 f''(\xi_1)/6$ and $R_2 = h^2 f''(\xi_2)/2$ for some points ξ_1 and ξ_2 between a and $a + h$. Therefore $R_1 - \frac{1}{2}hR_2 = h^3(f''(\xi_1)/6 - f''(\xi_2)/4)$. Since ξ_1 and ξ_2 need not be the same point, and, indeed, $f''(\xi_1)$ and $f''(\xi_2)$ need

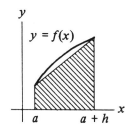

Fig. 10.4

not have the same sign, we cannot cancel and the best this approach gives is

$$|R_1 - \tfrac{1}{2}hR_2| \le h^3(|f''(\xi_1)|/6 + |f''(\xi_2)|/4) \le (5/12)Mh^3$$

where $M = \sup\{|f''(x)|: a \le x \le a+h\}$.

By the use of a spot of cunning, however, we can improve this. Write R_1 and R_2 in their integral forms (from Corollary 3 to Theorem 10.12):

$$R_1 = \tfrac{1}{2}\int_a^{a+h}(a+h-t)^2 f''(t)\,\mathrm{d}t, \quad R_2 = \int_a^{a+h}(a+h-t)f''(t)\,\mathrm{d}t$$

so that $R_1 - \tfrac{1}{2}hR_2 = \tfrac{1}{2}\int_a^{a+h}(a+h-t)(a-t)f''(t)\,\mathrm{d}t$.

Since, for $t \in [a, a+h]$, $a-t \le 0$ and $a+h-t \ge 0$,

$$|R_1 - \tfrac{1}{2}hR_2| \le \tfrac{1}{2}\int_a^{a+h}|a+h-t|\cdot|a-t|\cdot|f''(t)|\,\mathrm{d}t$$

$$\le \tfrac{1}{2}\int_a^{a+h}(a+h-t)(t-a)M\,\mathrm{d}t$$

$$= Mh^3/12.$$

Therefore, for a fixed function f and variable h, the error in this estimate decreases with h with order h^3 as $h \to 0$. We call this method the 'Trapezium Rule'. It can be used to evaluate integrals to any desired accuracy, even when an 'antiderivative' is not available. Suppose we wish to find $\int_0^1 f$, and that we use the Trapezium Rule with n subintervals each of length $1/n$. Then, if $M = \sup\{|f''(x)|: 0 \le x \le 1\}$ the error on each subinterval will not exceed $M/(12n^3)$ so the total error on adding the results for each subinterval will be at most $M/(12n^2)$, which obviously tends to 0 as $n \to \infty$.

It turns out that there is a more accurate approximation. Let $h > 0$ and consider the integral $\int_{a-h}^{a+h}f$. Letting $F(x) = \int_a^x f$ as before,

$$F(a+h) - F(a-h) = 2hf(a) + (h^3/3)f''(a) + R_1$$

where

$$R_1 = (1/24)\left\{\int_a^{a+h}(a+h-t)^4 f^{(4)}(t)\,\mathrm{d}t - \int_a^{a-h}(a-h-t)^4 f^{(4)}(t)\,\mathrm{d}t\right\},$$

obtained by using Taylor's Theorem on $[a, a+h]$ and $[a-h, a]$ separately. We wish to take a combination of $f(a-h)$, $f(a)$ and $f(a+h)$ which will give the terms in h and in h^3 above, with a remainder of order h^5. Now $f(a+h) + f(a-h) = 2f(a) + 2h^2 f''(a)/2 + R_2$ where

$$R_2 = (1/6)\left\{\int_a^{a+h}(a+h-t)^3 f^{(4)}(t)\,\mathrm{d}t + \int_a^{a-h}(a-h-t)^3 f^{(4)}(t)\,\mathrm{d}t\right\}.$$

Therefore

$$(h/3)\{f(a-h) + 4f(a) + f(a+h)\} = 2hf(a) + (h^3/3)f''(a) + (h/3)R_2$$

so that

$$\int_{a-h}^{a+h} f - (h/3)\{f(a-h) + 4f(a) + f(a+h)\} = R_1 - (h/3)R_2.$$

Now

$$R_1 - (h/3)R_2 = \int_a^{a+h} \{(a+h-t)^4/24 - (a+h-t)^3 h/18\} f^{(4)}(t)\,dt$$

$$- \int_a^{a-h} \{(a-h-t)^4/24 + (a-h-t)^3 h/18\} f^{(4)}(t)\,dt.$$

On noticing that

$$(a+h-t)^4/24 - (a+h-t)^3 h/18 = (a+h-t)^3(3(a-t)-h)/72 \le 0$$

for $t \in [a, a+h]$, we see that

$$\left| \int_a^{a+h} \{(a+h-t)^4/24 - (a+h-t)^3 h/18\} f^{(4)}(t)\,dt \right|$$

$$\le \int_a^{a+h} \{h(a+h-t)^3/18 - (a+h-t)^4/24\} |f^{(4)}(t)|\,dt$$

$$\le M_1 h^5/180$$

where $M_1 = \sup\{|f^{(4)}(t)| : a \le t \le a+h\}$. We obtain a similar estimate for the other integral, so, combining, we see that $|R_1 - (h/3)R_2| \le Mh^5/90$ where here $M = \sup\{|f^{(4)}(t)| : a-h \le t \le a+h\}$. This yields Simpson's Rule,

$$\left| \int_{a-h}^{a+h} f - (h/3)\{f(a-h) + 4f(a) + f(a+h)\} \right| \le Mh^5/90.$$

The result is interesting not only in its usefulness, as the 'error' tends to zero very rapidly as $h \to 0+$, but also as an illustration of how a little perseverance can extend the value of Taylor's Theorem. If we write the estimate for the integral as $2h\{f(a-h) + 4f(a) + f(a+h)\}/6$ it seems more plausible, being the length of the interval times a (complicated) average of the values of f at $a-h, a$ and $a+h$.

10.3   IMPROPER INTEGRALS

There are still some issues to be tackled. If we view the integral as the area under a curve, there is the possibility that this may exist for some curves defined on the whole real line, or on $[0, \infty)$. In this connection, we notice that $\int_0^a e^{-t}\,dt = 1 - e^{-a} \to 1$ as $a \to \infty$, giving an obvious candidate for the area under the curve $\{(t, e^t) : t \ge 0\}$. We make the definition which corresponds to this idea:

Definition   Let $f : [a, \infty) \to \mathbf{R}$ and suppose that for all $b > a$, f is integrable on $[a, b]$. If $\int_a^b f$ tends to a limit as $b \to \infty$, we say that the integral $\int_a^\infty f$ exists (or converges) and define $\int_a^\infty f = \lim_{b \to \infty} \int_a^b f$. The analogous definition is made for $\int_{-\infty}^a f$. If, for some $a \in \mathbf{R}$, both $\int_{-\infty}^a f$ and $\int_a^\infty f$ exist we say that $\int_{-\infty}^\infty f$ exists and has the value $\int_{-\infty}^a f + \int_a^\infty f$.

The existence of these 'improper' integrals is a matter broadly similar to that of the convergence of infinite series, and we shall develop tests to show that certain integrals exist, these being essential when a usable formula for $\int_a^b f$ does not exist. These 'improper' integrals occur in diverse places in mathematics and there are some powerful techniques for evaluating them, though we shall have to content ourselves with the question of existence.

Notice that the definition of $\int_{-\infty}^{\infty} f$ is equivalent to the existence of the limit of $\int_a^b f$ as $b \to \infty$ and $a \to -\infty$ *independently*. This is a stronger assertion than the existence of $\lim_{b \to \infty} \int_{-b}^b f$, as may be seen by considering $f(t) = t \;\forall t \in \mathbf{R}$; $\lim_{b \to \infty} \int_{-b}^b f = 0$ but neither $\int_0^{\infty} f$ nor $\int_{-\infty}^0 f$ exists. In this case, since $\int_0^a f$ exists for all $a \in \mathbf{R}$, it is easily seen that the non-existence of $\int_0^{\infty} f$ implies the non-existence of $\int_a^{\infty} f$ for all $a \in \mathbf{R}$.

It is worth noticing now that if $a \in \mathbf{R}$, $f: \mathbf{R} \to \mathbf{R}$ and both $\int_a^{\infty} f$ and $\int_{-\infty}^a f$ exist, then f is, by definition, integrable on $[a, b]$ for all $b \in \mathbf{R}$ so that $\int_b^{\infty} f$ and $\int_{-\infty}^b f$ will exist, as can be seen by noticing that for all x, $\int_a^x f = \int_a^b f + \int_b^x f$. The point a appearing in the definition of $\int_{-\infty}^{\infty} f$ is therefore immaterial.

We proceed by analogy with infinite series.

Lemma 10.15 (Comparison Test) Suppose that $\forall x \geq a$ $f(x) \geq 0$ and f is integrable on $[a, x]$. If there is a constant K such that $\forall x \geq a$ $\int_a^x f \leq K$, then $\int_a^{\infty} f$ exists. If $\forall x \geq a$ $0 \leq f(x) \leq g(x)$ and $\int_a^{\infty} g$ exists, then $\int_a^{\infty} f$ exists.

Proof: Let $F(x) = \int_a^x f$. If $\forall x \geq 0$ $f(x) \geq 0$ then F is increasing, since $y > x \Rightarrow F(y) - F(x) = \int_x^y f \geq 0$, and if F is bounded then $F(x) \to \sup F$ as $x \to \infty$. For the second part, if $\int_a^{\infty} g$ exists, then $\forall x \geq 0$ $0 \leq \int_a^x f \leq \int_a^x g \leq \int_a^{\infty} g$ so $\int_a^{\infty} f$ exists by the first part. $\square$

Example $\int_{-\infty}^{\infty} e^{-x^2} \, dx$ exists. Since e^{-x^2} is a continuous function of x, $\int_a^b e^{-x^2} \, dx$ exists for all finite a and b, so we need only consider the existence of $\int_1^{\infty} e^{-x^2} \, dx$ and $\int_{-\infty}^{-1} e^{-x^2} \, dx$. Now $x \geq 1 \Rightarrow x^2 \geq x \Rightarrow 0 \leq e^{-x^2} \leq e^{-x}$ so, since $\int_1^{\infty} e^{-x} \, dx \leq 1$, $\int_1^{\infty} e^{-x^2} \, dx$ exists by comparison. The existence of $\int_{-\infty}^{-1} e^{-x^2} \, dx$ is similar.

Lemma 10.16 (Absolute Convergence) Suppose that for all $x \geq a$, f is integrable on $[a, x]$ and that $\int_a^{\infty} |f|$ exists. Then $\int_a^{\infty} f$ exists.

Proof: Let $f_+(x) = \max(f(x), 0) = \frac{1}{2}(f(x) + |f(x)|)$, and $f_-(x) = \max(-f(x), 0)$. For all $x \geq a$, $f_{\pm}$ are integrable on $[a, x]$ and $0 \leq f_{\pm}(x) \leq |f(x)|$. By comparison $\int_a^{\infty} f_{\pm}$ both exist, hence so does $\int_a^{\infty} f = \int_a^{\infty} f_+ - \int_a^{\infty} f_-$. $\square$

Lemma 10.17 (The Integral Test) Suppose that $f: [1, \infty] \to \mathbf{R}$ is non-negative and decreasing; then $\int_1^{\infty} f$ exists if and only if $\sum f(n)$ converges.

Proof: $n \leq x \leq n+1 \Rightarrow f(n) \geq f(x) \geq f(n+1)$, so $f(n) \geq \int_n^{n+1} f \geq f(n+1)$, this being true for all $n \in \mathbf{N}$. Letting $s_n = f(1) + f(2) + \cdots + f(n)$, we see that $\forall n \geq 2$ $s_{n-1} \geq \int_1^n f \geq s_n - f(1)$. This double inequality shows us that (s_n) is a bounded sequence if and only if $(\int_1^n f)$ is a bounded sequence. Thus if $\int_1^{\infty} f$

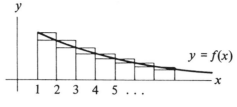

y

$y = f(x)$

1 2 3 4 5 ...

x

Fig. 10.5

exists, $(\int_1^n f)$ is a bounded sequence, so (s_n) is increasing and bounded above, hence convergent. Conversely, if $\sum f(n)$ converges, $(\int_1^n f)$ is a bounded sequence, so if K is an upper bound for $(\int_1^n f)$, we see that if $x \geq 1$, $\exists n \in \mathbf{N}$ s.t. $n > x$ whence $\int_1^x f \leq \int_1^n f \leq K$. This shows that $\forall x \geq 1$ $\int_1^x f \leq K$ and $\int_1^\infty f$ exists by Lemma 10.15. □

As a convergence test for series, this gives us more or less the same information as the condensation test, though it is more transparent. We see, for example, that $\int_1^\infty (1/x^\alpha)\,dx$ exists if and only if $\alpha > 1$, either by direct calculation or by the knowledge that $\sum (1/n^\alpha)$ converges if and only if $\alpha > 1$. The technique of the integral test is, however, more useful than this, since it gives us estimates of the difference between $\sum_{n=1}^\infty f(n)$ and $\sum_{n=1}^N f(n)$ in some cases.

With the above notation, $\forall n > M$ $\sum_{n=M}^{N-1} f(n) \geq \int_M^N f \geq \sum_{n=M+1}^N f(n)$ whence, letting $N \to \infty$,

$$\sum_{n=M}^\infty f(n) \geq \int_M^\infty f \geq \sum_{n=M+1}^\infty f(n)$$

(when the series and the integral converge). Let $f(x) = 1/x^2$ so that if $s_n = 1 + 1/2^2 + \cdots + 1/n^2$, and $s = \sum_{n=1}^\infty 1/n^2$,

$$s - s_M = \sum_{n=M+1}^\infty 1/n^2 \leq \int_M^\infty f = \frac{1}{M}, \quad \text{and} \quad s - s_M \geq \int_{M+1}^\infty f = \frac{1}{M+1}$$

so $1/(M+1) \leq s - s_M \leq 1/M$, giving a very simple estimate of the difference between the sum to M terms and the sum to infinity.

In practice, the convergence of most improper integrals is established by comparison, with the occasional assistance of manipulative devices like integration by parts.

Example $\int_0^\infty (\sin x)/x\,dx$ converges. First notice that $(\sin x)/x \to 1$ as $x \to 0+$, so if we give the integrand the value 1 at $x = 0$, the resulting function is continuous, and $\int_0^a (\sin x)/x\,dx$ exists for all $a > 0$. It is therefore enough to show the existence of $\int_1^\infty (\sin x)/x\,dx$.

At first sight, one might notice that $x > 0 \Rightarrow |(\sin x)/x| \leq 1/x$ and try the comparison test, but this is fruitless since $\int_1^\infty 1/x\,dx$ does not exist. Let $F(y) = \int_1^y (\sin x)/x\,dx$. Then, integrating by parts,

$$F(y) = \cos 1 - (\cos y)/y - \int_1^y (\cos x)/x^2\,dx.$$

Since $\int_1^\infty 1/x^2\,dx$ does exist and $x\geq 1\Rightarrow|(\cos x)/x^2|\leq 1/x^2$ we see that $\int_1^\infty(\cos x)/x^2\,dx$ exists. Therefore the last term on the right of the expression for $F(y)$ tends to a limit as $y\to\infty$, whence, since $(\cos y)/y\to 0$ as $y\to\infty$, $F(y)$ tends to a limit as $y\to\infty$.

There is another type of improper integral. In the definition of $\int_a^b f$ we demanded that f be a bounded function, yet we can ascribe a value to such integrals, in certain cases where f is unbounded, in much the same way as to $\int_a^\infty f$.

Definition Suppose that $f:(a,b]\to\mathbf{R}$ and that for all $x\in(a,b]$ f is integrable on $[x,b]$. Then we say that the (improper) integral $\int_a^b f$ exists if $\int_x^b f$ tends to a limit as $x\to a+$. The analogous definition applies if $\lim_{x\to b-}\int_a^x f$ exists. If $f:[a,b]\setminus\{c\}\to\mathbf{R}$ and both $\int_a^c f$ and $\int_c^b f$ exist, we say the (improper) integral $\int_a^b f$ exists and has the value $\int_a^b f=\int_a^c f+\int_c^b f$.

Example $\int_0^1 x^{-\alpha}\,dx$ exists iff $\alpha<1$. For $\int_a^1 x^{-\alpha}\,dx=(1-a^{1-\alpha})/(1-\alpha)\to 1/(1-\alpha)$ as $a\to 0+$ if $1-\alpha>0$, i.e. $\alpha<1$, while the limit does not exist if $\alpha>1$. The limit is also non-existent if $\alpha=1$ since $\int_a^1 x^{-1}\,dx=-\log a$.

The following two analogues of the results for $\int_a^\infty f$ are proved by simple changes to the proofs of Lemmas 10.15 and 10.16.

Lemma 10.18 (Comparison Test) Suppose that $\forall x\in(a,b]$ $f(x)\geq 0$ and f is integrable on $[x,b]$. If $\{\int_x^b f:x\in(a,b]\}$ is bounded then $\int_a^b f$ exists. If $\int_a^b g$ exists and $\forall x\in(a,b]$ $0\leq f(x)\leq g(x)$ then $\int_a^b f$ exists.

Lemma 10.19 (Absolute Convergence) Suppose that $\forall x\in(a,b]$ f is integrable on $[x,b]$ and that $\int_a^b|f|$ exists. Then $\int_a^b f$ exists.

The two types of integral may be combined. For example, if we wished to examine $\int_0^\infty dx/\sqrt{(x\cdot|x-1|\cdot|x-2|)}$ we would investigate, separately, the existence of the improper integrals on $(0,\frac{1}{2}]$, $[\frac{1}{2},1)$, $(1,\frac{3}{2}]$, $[\frac{3}{2},2)$, $(2,3]$ and $[3,\infty)$. On $(0,\frac{1}{2}]$ the function is continuous so the only issue is the behaviour near 0. Since $1/\sqrt{(|x-1|\cdot|x-2|)}$ is continuous on $[0,\frac{1}{2}]$ it is bounded there; let M be an upper bound. Then $\forall x\in(0,\frac{1}{2}]$, $1/\sqrt{(x|x-1||x-2|)}\leq Mx^{-\frac{1}{2}}$ so $\int_0^{\frac{1}{2}}$ exists by comparison. The existence on the other four bounded intervals is similarly proved, while on $[3,\infty)$ we see that $x\geq 3\Rightarrow x(x-1)(x-2)\geq(2/9)x^3\Rightarrow 1/\sqrt{(x|x-1|\,|x-2|)}\leq\sqrt{(9/2)}x^{-3/2}$, so $\int_3^\infty$ exists by comparison with $\int_3^\infty x^{-3/2}\,dx$.

Example The Gamma Function, $\Gamma:(0,\infty)\to\mathbf{R}$, is defined as follows:

$$\Gamma(x)=\int_0^\infty t^{x-1}e^{-t}\,dt.$$

The integral exists if both $\int_0^1$ and $\int_1^\infty$ exist. The former obviously exists if $x\geq 1$ for then the integrand is continuous on $[0,1]$. If $x-1<0$ the integrand is unbounded near $t=0$, so since $0<t\leq 1$ implies $0<t^{x-1}e^{-t}<t^{x-1}$, and $\int_0^1 t^{x-1}\,dt$ converges iff $x-1>-1$, we see that $\int_0^1 t^{x-1}e^{-t}\,dt$ exists for $x>0$.

(This integral diverges by the comparison test for $x < 0$ since $0 < t \leq 1 \Rightarrow$ $t^{x-1} e^{-t} \geq t^{x-1} e^{-1}$.) Since $t^\alpha e^{-\lambda t} \to 0$ as $t \to \infty$ for $\lambda > 0$, we see that $t^{x-1} e^{-t/2} \to 0$ as $t \to \infty$ so there is a constant T such that $t \geq 1 \Rightarrow$ $t^{x-1} e^{-t/2} \leq T$. Then $\forall t \geq 1 \; 0 \leq t^{x-1} e^{-t} \leq T e^{-t/2}$ so $\int_1^\infty t^{x-1} e^{-t} \, dt$ converges by comparison. Now, for $x > 0$,

$$\Gamma(x+1) = \int_0^\infty t^x e^{-t} \, dt = \lim_{\substack{b \to \infty \\ a \to 0+}} \int_a^b t^x e^{-t} \, dt$$

$$= \lim_{\substack{b \to \infty \\ a \to 0+}} \left\{ -b^x e^{-b} + a^x e^{-a} + \int_a^b x t^{x-1} e^{-t} \, dt \right\}$$

$$= x \Gamma(x)$$

since $b^x e^{-b} \to 0$ as $b \to \infty$ and $a^x e^{-a} \to 0$ as $a \to 0+ \; (x > 0)$. Therefore $\forall x > 0$ $\Gamma(x+1) = x\Gamma(x)$. Moreover, $\Gamma(1) = 1$ so it is easily seen that $\forall n \in \mathbf{N} \; \Gamma(n) = (n-1)!$ and we have found a function defined on $\mathbf{R}$ whose value at the natural numbers is a factorial. At this point we recall the formula for the nth indefinite integral of the continuous function f. Rewriting the $(n-1)!$ as $\Gamma(n)$ we obtain

$$F_n(x) = \frac{1}{\Gamma(n)} \int_0^x (x-t)^{n-1} f(t) \, dt,$$

an expression which has a meaning even if n is not an integer! This apparently whimsical observation can be put to good use in more sophisticated theories of differential equations.

Problems

1. Let $f(x) = x^2$ for $0 \leq x \leq 1$ and let $D_n = \{0, 1/n, \ldots, (n-1)/n, 1\}$ be the dissection of $[0, 1]$ into n subintervals of equal length. Calculate $S(D_n)$ and show that $\inf \{S(D_n) : n \in \mathbf{N}\} = \sup \{s(D_n) : n \in \mathbf{N}\} = 1/3$. Deduce that $\int_0^1 x^2 \, dx = 1/3$.
    [You may need to recall that $\sum_{k=1}^n k^2 = n(n+1)(2n+1)/6$.]

2. Let $a < b < c$ and define $f : [a, c] \to \mathbf{R}$ by $f(x) = 1$ if $a \leq x \leq b$ and $f(x) = 0$ if $b < x \leq c$. By calculating $\inf \{S(D)\}$ and $\sup \{s(D)\}$ directly, show that $\int_a^c f = b - a$.

3. Suppose that $f : [a, b] \to \mathbf{R}$ is integrable and that $\forall x \in [a, b] \; f(x) \geq 0$. By Lemma 10.10, then, $\int_a^b f \geq 0$. Show that if, for some $\alpha > 0$ and $\delta > 0$, there is an interval $(c - \delta, c + \delta) \subset [a, b]$ for which $\forall x \in (c - \delta, c + \delta)$ $f(x) \geq \alpha$ then $\int_a^b f > 0$.

4. Let $f : [a, b] \to \mathbf{R}$ be continuous and satisfy $\forall x \in [a, b] \; f(x) \geq 0$. Show, using Q3, that if $\exists c \in [a, b]$ for which $f(c) > 0$ then $\int_a^b f > 0$. Deduce that if g is continuous and $\int_a^b |g| = 0$ then g is identically zero (i.e. $\forall x \in [a, b]$ $g(x) = 0$).

5. Suppose that $f : [a, b] \to \mathbf{R}$ is continuous and that for all continuous functions $\phi : [a, b] \to \mathbf{R}$, $\int_a^b f\phi = 0$. Show that f is identically 0.

6. Let $f:[a,b] \to \mathbf{R}$ be integrable and define F by $F(x) = \int_a^x f$. Prove that F is continuous, and deduce that there is a point $\xi \in [a,b]$ for which $\int_a^\xi f = \int_\xi^b f$.

7. Let $f:[0,a] \to \mathbf{R}$ be continuous. Show that $\lim_{x \to 0+} (1/x)\int_0^x f(t)\,dt = f(0)$.

8. By showing that the theorem for differentiation under the integral sign applies, show that if $y(x) = \int_0^x \sin(x-t)f(t)\,dt$, where f is a given continuous function, then y satisfies the differential equation $y'' + y = f$ and $y(0) = y'(0) = 0$.

9. By showing that, for $x > 1$, $\int_1^x (1/t)\,dt = \log x$, prove that $\forall n \in \mathbf{N}$ $1/(n+1) \le \log(1 + 1/n) \le 1/n$. Let $s_n = 1 + 1/2 + 1/3 + \cdots + 1/n$ and $a_n = s_n - \log n$. Show that (a_n) is decreasing and that $\forall n \in \mathbf{N}$ $1/n \le a_n \le 1$. Deduce that there is a number $\gamma \in [0,1]$ such that $a_n \to \gamma$ as $n \to \infty$, and that $s_n = \log n + \gamma + \varepsilon_n$ where $0 \le \varepsilon_n \le 1$ and $\varepsilon_n \to 0$ as $n \to \infty$.

10. Use the Trapezium Rule with no intermediate steps on $[n, n+1]$ to show that $\int_n^{n+1} \log t\,dt = (\log n + \log(n+1))/2 + \varepsilon_n$ where $|\varepsilon_n| \le 1/(12n^2)$. By calculating $\int_1^n \log t\,dt$ exactly (integrate by parts!), show that

$$\log(n!) = (n + \tfrac{1}{2})\log n - n + \phi(n)$$

where $\phi(n)$ tends to a limit as $n \to \infty$. Deduce that $n!/((n/e)^n\sqrt{n})$ tends to a limit as $n \to \infty$.

11. Let $I_n = \int_0^{\pi/2}(\sin t)^n\,dt$ $(n = 0, 1, 2, \ldots)$. Show that for all $n \ge 0$, $I_n \ge I_{n+1} > 0$ and that if $n \ge 2$, $I_n = ((n-1)/n)I_{n-2}$. Deduce that $I_{2n} \ge I_{2n+1} \ge ((2n+1)/(2n+2))I_{2n}$ and hence that $I_{2n+1}/I_{2n} \to 1$ as $n \to \infty$. By using the expression $I_n = ((n-1)/n)I_{n-2}$ show that

$$I_{2n+1} = \frac{(2n)(2n-2)\ldots 4.2}{(2n+1)(2n-1)\ldots 5.3} = \frac{(2^n n!)^2}{(2n+1)!} \quad \text{and} \quad I_{2n} = \frac{(2n)!}{(2^n n!)^2}\frac{\pi}{2}.$$

Deduce that

$$\lim_{n \to \infty} \frac{1}{\sqrt{(2n+1)}}\frac{(2^n n!)^2}{(2n)!} = \sqrt{\frac{\pi}{2}}.$$

Use this to show that $\lim_{n \to \infty} n!/(\sqrt{n}(n/e)^n) = \sqrt{(2\pi)}$ (cf. Q10); hence show that the ratio of $n!$ to $\sqrt{(2\pi n)}\cdot(n/e)^n$ tends to 1 as $n \to \infty$.

12. Show that the following improper integrals converge:

$$\int_0^\infty \frac{dx}{1+x^2}, \quad \int_0^\infty e^{-ax}\sin(bx)\,dx \quad (a > 0), \quad \int_1^\infty \frac{\sin x}{x^\alpha}\,dx \quad (\alpha > 0),$$

$$\int_0^1 \log x\,dx, \quad \int_0^1 \frac{dx}{\sqrt{(1-x)}}, \quad \int_0^1 \frac{dx}{\sqrt{(x(1-x))}}, \quad \int_{-\infty}^\infty e^{-|x|}\,dx.$$

13. Suppose that $f:[0,\infty) \to \mathbf{R}$ and that $\int_0^\infty |f|$ exists. Show that, for all $\xi \in \mathbf{R}$, $\int_0^\infty f(x)\sin(\xi x)\,dx$ and $\int_0^\infty f(x)\cos(\xi x)\,dx$ both converge.

14. Show that $\int_0^\infty (\cos x)/\sqrt{x}\,dx$ converges and deduce that $\int_0^\infty \cos(t^2)\,dt$ converges. This is noteworthy since $\cos(t^2) \nrightarrow 0$ as $t \to \infty$.

15. For $\xi \in \mathbf{R}$ define $F(\xi) = \int_0^\infty (\sin(\xi x))/x\,dx$. Show that if $\xi > 0$ then $F(\xi) = F(1)$ and if $\xi < 0$ then $F(\xi) = -F(1)$. It follows that if we can show $F(1) > 0$, F must be discontinuous at 0.

16.* Show that $\int_0^\infty (\sin x)/x\,dx = \sum_{n=1}^\infty \int_{2(n-1)\pi}^{2n\pi} (\sin x)/x\,dx$. By splitting the range of integration into two, or otherwise, show that $\int_{2(n-1)\pi}^{2n\pi} (\sin x)/x\,dx > 0$ and hence that $\int_0^\infty (\sin x)/x\,dx > 0$.

17. Which of the following improper integrals exist?

$$\int_0^\infty \frac{dx}{x^2 - 1}, \qquad \int_0^\infty \frac{dx}{(x+1)\sqrt{|x^2 - 1|}}, \qquad \int_0^\infty t\,e^{-t}\,dt.$$

For which real α does $\int_0^\infty x^\alpha\,dx$ exist?

18. Show that $\sum_{n \geq 2} 1/(n(\log n)^2)$ converges and show that if s_n denotes the sum up to the term $1/(n(\log n)^2)$ inclusive, and s is the sum to infinity, then $1/\log(n+1) < s - s_n < 1/\log n$. Deduce that for the partial sum s_n to be 'correct to two decimal places' (i.e. within $1/200$ of the sum to infinity) we must let n be at least $e^{200} - 1$.

19.* Suppose that $f:[a,b] \to \mathbf{R}$ is bounded and that for all $x \in (a,b]$ f is integrable on $[x,b]$. Show that $\forall \varepsilon > 0$ there is a dissection D of $[a,b]$ with $S(D) - s(D) < \varepsilon$, and hence that f is integrable. Deduce that $\int_0^1 \sin(1/x)\,dx$ exists.

20.* (i) Prove that if $c < d$, $\int_c^d \sin(\lambda x)\,dx \to 0$ as $\lambda \to \infty$.

(ii) $g:[a,b] \to \mathbf{R}$ is said to be a 'step function' if there are points $c_0, \ldots, c_n$ with $a = c_0 < c_1 < \cdots < c_n = b$ such that g is constant on the intervals (c_{i-1}, c_i) and the value of $g(c_i)$ is that taken by g on one of the adjacent intervals. Prove that if g is a step function then

$$\forall \varepsilon > 0 \quad \exists \Lambda \quad \text{s.t.} \quad \forall \lambda \geq \Lambda \quad \left| \int_a^b g(x)\sin(\lambda x)\,dx \right| < \tfrac{1}{2}\varepsilon.$$

(iii) Let $f:[a,b] \to \mathbf{R}$ be integrable. Show directly from the definition of integrability that there is a step function g on $[a,b]$ for which $\forall x \in [a,b]\ g(x) \leq f(x)$ and $\int_a^b (f - g) < \tfrac{1}{2}\varepsilon$. (What is a lower sum?) Deduce that there is a step function g such that $\int_a^b |f - g| < \tfrac{1}{2}\varepsilon$.

(iv) Assemble all this into a proof that

$$\int_a^b f(x)\sin(\lambda x)\,dx \to 0 \quad \text{as} \quad \lambda \to \infty.$$

21.* Suppose that f is integrable on $[a,b]$ and that $\forall x \in [a,b]\ f(x) > 0$. We wish to show $\int_a^b f > 0$, which is rather more awkward than one might expect.

Assume the result is false, i.e. that $\int_a^b f = 0$.

(i) Show that if $[c,d] \subset [a,b]$ then $\int_c^d f = 0$.

(ii) Prove that there is a dissection D_1 of $[a,b]$ with $S(D_1) \leq b - a$ and deduce that there is a subinterval $[a_1, b_1]$ of $[a, b]$ such that $a_1 < b_1$ and $\sup_{[a_1, b_1]} f \leq 1$.

(iii) By noticing that $\int_{a_1}^{b_1} f = 0$, repeat the process to show by induction that there is a sequence of intervals $[a_n, b_n]$ such that $\forall n \in \mathbf{N}$ $a_n < b_n$, $[a_{n+1}, b_{n+1}] \subset [a_n, b_n]$ and $\sup_{[a_n, b_n]} f \leq 1/n$.

(iv) Show that (a_n) is increasing and bounded above, and let $x = \lim_{n \to \infty} a_n$.

Show that $\forall n \in \mathbf{N}$ $x \in [a_n, b_n]$, and deduce that $f(x) = 0$. Use this contradiction to show that $\int_a^b f > 0$.

CHAPTER 11

Functions of Several Variables

"And thick and fast they came at last, and
more, and more, and more."

Lewis Carroll

There are evidently many mathematical quantities which depend on more than one variable so we must augment our analysis to include the consideration of functions of several variables. Much of what we have established for functions of one variable remains true in the new situation, but some features change; to oversimplify, the ε–δ techniques remain much the same, the extra variables adding a certain long-windedness, but some under lying features change, this arising from the fact that the geometry of n-dimensional space is rather different from that of one dimension.

The first issue is to decide what we mean by limits and continuity. We denote by $\mathbf{R}^n$ the set of all n-tuples $(x_1, \ldots, x_n)$ of real numbers; where helpful, we shall abbreviate $(x_1, \ldots, x_n)$ as $\mathbf{x}$. To define a limit as $\mathbf{x}$ tends to $\mathbf{a}$ we need an idea of distance, which we generalise from that used in two- and three-dimensional geometry. The definitions of limit and continuity are then obtained from the idea that $f(\mathbf{x}) \to L$ as $\mathbf{x} \to \mathbf{a}$ if $|f(\mathbf{x}) - L|$ can be made as small as desired by choosing $\mathbf{x}$ 'near enough' to $\mathbf{a}$.

Definitions   Let $\mathbf{x} = (x_1, \ldots, x_n) \in \mathbf{R}^n$. The **norm** of $\mathbf{x}$, denoted by $\|\mathbf{x}\|$, is the quantity $\|\mathbf{x}\| = \sqrt{\sum_{i=1}^n |x_i|^2}$. The distance apart of two points $\mathbf{x}$ and $\mathbf{y}$ is said to be $\|\mathbf{x} - \mathbf{y}\|$, where $\mathbf{x} - \mathbf{y} = (x_1 - y_1, \ldots, x_n - y_n)$.

Let $\mathbf{a} = (a_1, \ldots, a_n)$ and suppose that $f : A \to \mathbf{R}$ where A is some subset of $\mathbf{R}^n$ containing $\{\mathbf{x} \in \mathbf{R}^n : 0 < \|\mathbf{x} - \mathbf{a}\| < R\}$ for some $R > 0$. We say $f(\mathbf{x}) \to L$ as $\mathbf{x} \to \mathbf{a}$ if and only if

$$\forall \varepsilon > 0 \quad \exists \delta > 0 \quad \text{s.t.} \quad \forall \mathbf{x} \in \mathbf{R}^n \quad 0 < \|\mathbf{x} - \mathbf{a}\| < \delta \Rightarrow |f(\mathbf{x}) - f(\mathbf{a})| < \varepsilon.$$

Suppose that $A \subset \mathbf{R}^n$ and that $f : A \to \mathbf{R}$. Then if $\mathbf{a} \in A$ we say that f is continuous at $\mathbf{a}$ if and only if

$$\forall \varepsilon > 0 \quad \exists \delta > 0 \quad \text{s.t.} \quad \forall \mathbf{x} \in A \quad \|\mathbf{x} - \mathbf{a}\| < \delta \Rightarrow |f(\mathbf{x}) - f(\mathbf{a})| < \varepsilon.$$

f is said to be continuous if it is continuous at each point of A. If $\mathbf{a}$ is 'interior to' A in the sense that there is a positive h for which $\{\mathbf{x} : \|\mathbf{x} - \mathbf{a}\| < h\} \subset A$, then continuity at $\mathbf{a}$ is exactly the same as the property that $f(\mathbf{x}) \to f(\mathbf{a})$ as $\mathbf{x} \to \mathbf{a}$. If, however, $\mathbf{a} \in A$ but $\mathbf{a}$ is not interior to A then continuity at $\mathbf{a}$ does not imply $f(\mathbf{x}) \to f(\mathbf{a})$ as $\mathbf{x} \to \mathbf{a}$, since the definition of continuity only pays heed to those points $\mathbf{x}$ near $\mathbf{a}$ which happen to lie in A. This is akin to the definition of a continuous function on the interval $[0, 1]$ where we make a slightly less stringent demand for continuity at the endpoints. Since the geometry of the 'edge' of a set in $\mathbf{R}^n$ is more complicated than that in one dimension, we cannot just formulate what we wish in terms of one-sided continuity. It is easy to see that if A is an interval in $\mathbf{R}^1$, the new definition of a continuous function from A to $\mathbf{R}$ coincides with our old one, though the new one also allows us to consider less regular sets than intervals.

Examples   Let $f:\mathbf{R}^2 \to \mathbf{R}$ be given by $f(x, y) = xy$. (We shall label the variables x and y in place of x_1 and x_2 to accord with tradition.) Then $f(x, y) \to 1$ as $(x, y) \to (1, 1)$.

Let $\varepsilon > 0$. Let $\delta = \min(1, \varepsilon/3)$.

Then $\|(x, y) - (1, 1)\| < \delta \Rightarrow \sqrt{((x - 1)^2 + (y - 1)^2)} < \delta$

$$\Rightarrow |x - 1| < \delta \quad \text{and} \quad |y - 1| < \delta$$
$$\Rightarrow |x| < 2 \quad \text{and} \quad |y| < 2 \text{ (since } \delta \leq 1\text{)}.$$

Therefore $\|(x, y) - (1, 1)\| < \delta \Rightarrow |f(x, y) - 1| = |xy - 1|$

$$\leq |(x - 1)y| + |y - 1|$$
$$< 2\delta + \delta \leq \varepsilon.$$

Since $\varepsilon > 0$ was arbitrary we have shown the result.

From this it follows that f is continuous at $(1, 1)$ and a similar argument shows f is continuous everywhere.

More generally, let $g, h: A \to \mathbf{R}$ be continuous, where A is an interval in $\mathbf{R}$. Define $f: A \times A \to \mathbf{R}$ by $f(x, y) = g(x)h(y)$; f is continuous. To show this, let $(a, b) \in A \times A$ and let $\varepsilon > 0$.

Since g and h are continuous, $\exists \delta_1 > 0$ s.t. $\forall x \in A \ |x - a| < \delta_1 \Rightarrow |g(x) - g(a)| < \varepsilon$ and $\exists \delta_2 > 0$ s.t. $\forall y \in A \ |y - b| < \delta_2 \Rightarrow |h(y) - h(b)| < \varepsilon$. Also, $\exists \delta_3 > 0$ s.t. $\forall y \in A \ |y - b| < \delta_3 \Rightarrow |h(y) - h(b)| < 1$.

Then $\|(x, y) - (a, b)\| < \delta = \min(\delta_1, \delta_2, \delta_3) \Rightarrow |x - a| \leq \|(x, y) - (a, b)\| < \delta$ and $|y - b| \leq \|(x, y) - (a, b)\| < \delta$, so

$$\|(x, y) - (a, b)\| < \delta \Rightarrow |f(x, y) - f(a, b)| = |g(x)h(y) - g(a)h(b)|$$
$$\leq |g(x) - g(a)| \, |h(y)|$$
$$+ |h(y) - h(b)| \, |g(a)|$$
$$< \varepsilon(|h(y)| + |g(a)|)$$
$$< \varepsilon(1 + |h(b)| + |g(a)|).$$

Since the factor multiplying ε is a constant and the above is true for all $\varepsilon > 0$, we could have substituted $\varepsilon/(1 + |h(b)| + |g(a)|)$ for ε throughout; this shows f is continuous at (a, b) and since (a, b) was typical, f is continuous.

Lemma 11.1   Let $f, g: A \to \mathbf{R}$ be continuous, where $A \subset \mathbf{R}^n$, and let $\lambda \in \mathbf{R}$. Then $f + g, fg, \lambda f$ and $|f|$ are all continuous, and $1/f$ is continuous on the set $A \backslash \{\mathbf{x}: f(\mathbf{x}) = 0\}$.

We leave the proof of the Lemma as an exercise. From this and the example preceding it, we see that functions which can be defined by adding, multiplying and dividing functions of one variable are continuous except where their denominator is zero. Thus, for example, $(x_1 x_2 + x_3)/(x_1^2 + x_2^2 + x_3^2)$ is a continuous function of (x_1, x_2, x_3) except possibly at $(0, 0, 0)$.

Suppose that $f: \mathbf{R}^n \to \mathbf{R}$, so that $f(x_1, \ldots, x_n)$ is a function of the n independent variables $x_1, \ldots, x_n$. By fixing some of these variables we can obtain a related function dependent on fewer variables, which raises the question of how we relate the continuity of the various functions so obtained. Suppose $a_2, \ldots, a_n \in \mathbf{R}$ and $g_1: \mathbf{R} \to \mathbf{R}$ is defined by $g_1(x) = f(x, a_2, \ldots, a_n)$. Then if f is continuous at $(a_1, \ldots, a_n)$, g_1 is continuous at a_1. To see this, let $\varepsilon > 0$; $\exists \delta > 0$ s.t. $\|\mathbf{x} - \mathbf{a}\| < \delta \Rightarrow |f(\mathbf{x}) - f(\mathbf{a})| < \varepsilon$ whence

$$|x - a_1| < \delta \Rightarrow \|(x, a_2, \ldots, a_n) - (a_1, \ldots, a_n)\| < \delta \Rightarrow |g_1(x) - g_1(a_1)| < \varepsilon.$$

Similarly g_i, given by $g_i(x) = f(a_1, \ldots, a_{i-1}, x, a_{i+1}, \ldots, a_n)$, is continuous. This result we can paraphrase as saying that if f is continuous in the n variables $x_1, \ldots, x_n$ jointly, then f is continuous in each variable separately (i.e. fixing all but one of the variables). The converse result is false.

Example   Define $f: \mathbf{R}^2 \to \mathbf{R}$ by $f(x, y) = 2xy/(x^2 + y^2)$ (for $(x, y) \neq (0, 0)$) and $f(0, 0) = 0$. f is continuous at all points other than $(0, 0)$, since at all such points, $x^2 + y^2 \neq 0$. Now for $x \neq 0$, $f(x, x) - f(0, 0) = 1$ while $f(x, -x) - f(0, 0) = -1$, so, whatever $\delta > 0$ we choose, there are points (x, y) with $\|(x, y)\| < \delta$ for which $|f(x, y) - f(0, 0)| = 1$, so f is not continuous at $(0, 0)$—nor could a different choice of $f(0, 0)$ produce a continuous function.

The result here is more easily visualised by changing to polar coordinates and letting $x = r \cos \theta$, $y = r \sin \theta$ so $f(x, y) - f(0, 0) = 2 \cos \theta \sin \theta$. This does not tend to 0 as $r = \|(x, y)\| \to 0$ (unless $\cos \theta \sin \theta = 0$).

However, $f(x, 0) = 0$ and $f(0, y) = 0$ so the functions obtained from f by fixing one of the variables are continuous. We conclude that 'joint' continuity with respect to several variables is stronger than continuity with respect to each variable separately. This difference can be significant and occasionally awkward. The reader may take heart from the knowledge that, in the development of analysis, even such an eminent mathematician as Cauchy failed to notice this!

We shall prove an analogue of Theorem 7.4 which shows that if $f: [a, b] \to \mathbf{R}$ is continuous, it is bounded. For this we need to find a suitable type of set to form the domain of our function.

Definitions   Let $(\mathbf{x}_n)$ be a sequence of points of $\mathbf{R}^m$. We say that $\mathbf{x}_n \to \mathbf{x}$ as $n \to \infty$ if and only if $\forall \varepsilon > 0 \; \exists N$ s.t. $\forall n \geq N \; \|\mathbf{x}_n - \mathbf{x}\| < \varepsilon$. Let $A \subset \mathbf{R}^m$. We say A is **closed** if, whenever $(\mathbf{x}_n)$ is a sequence of elements of A and $\mathbf{x}_n \to \mathbf{x}$, then $\mathbf{x} \in A$.

Examples $\mathbf{R}^m$ itself is closed. Also $[a, b]$ is a closed subset of $\mathbf{R}^1$ (since if $\forall n, x_n \in [a, b]$ and $x_n \to x$ as $n \to \infty$, then $x \in [a, b]$). The set $(0, 1)$ is not closed since $\forall n \in \mathbf{N} \; 1/n \in (0, 1)$ and $1/n \to 0$ but $0 \notin (0, 1)$.

Lemma 11.2   Let A be a subset of $\mathbf{R}^m$ and $f: A \to \mathbf{R}$ be continuous. If $\forall n \in \mathbf{N}$ $\mathbf{x}_n \in A$ and $\mathbf{x}_n \to \mathbf{a} \in A$ as $n \to \infty$ then $f(\mathbf{x}_n) \to f(\mathbf{a})$ as $n \to \infty$.

This result, whose proof is a direct adaptation of that for sequences and functions defined in **R**, leads us to a useful technique for spotting closed sets.

Let $f: \mathbf{R}^m \to \mathbf{R}$ be continuous and $A = \{\mathbf{x}: f(\mathbf{x}) \in [a, b]\} = f^{-1}([a, b])$; A is closed. Notice that f has the whole of $\mathbf{R}^m$ as its domain. The proof is easy: let $\mathbf{x} \in \mathbf{R}^m$ and suppose that $\forall n \in \mathbf{N} \ \mathbf{x}_n \in A$ and $\mathbf{x}_n \to \mathbf{x}$. Then $f(\mathbf{x}_n) \to f(\mathbf{x})$ as $n \to \infty$ so, since $a \leq f(\mathbf{x}_n) \leq b$ for all n, $a \leq f(\mathbf{x}) \leq b$ and $\mathbf{x} \in A$. A is thus closed.

The same argument shows that $f^{-1}([a, \infty))$ and $f^{-1}((-\infty, a])$ are closed, and a slight change shows that the intersection of two closed sets, or of any collection of closed sets, is closed. Thus we can see easily that, for example, the 'disc' $\{\mathbf{x} \in \mathbf{R}^m: \|\mathbf{x}\| \leq 1\}$ is closed, since it is $\{\mathbf{x}: f(\mathbf{x}) \in [0, 1]\}$ where $f(\mathbf{x}) = \sqrt{(x_1^2 + \cdots + x_m^2)}$ and f is continuous. A useful intuitive view of a closed set is one which includes the points 'at its edges'; $\{\mathbf{x} \in \mathbf{R}^m: \|\mathbf{x}\| < 1\}$ is not closed but $\{\mathbf{x} \in \mathbf{R}^m: \|\mathbf{x}\| \leq 1\}$ is closed.

To proceed further, we need a short discussion on sequences.

Definition   Let (a_n) be a sequence of real numbers. A **subsequence** of (a_n) is a sequence consisting of some of the terms of (a_n) in their original order, that is, a sequence of the form $(a_{n_k})_{k=1}^{\infty}$ where $\forall k \in \mathbf{N} \ n_{k+1} > n_k$ and $n_k \in \mathbf{N}$. Thus $(a_2, a_4, a_6, \ldots)$ is a subsequence of (a_n).

Lemma 11.3   Every sequence of real numbers has either an increasing subsequence or a decreasing one.

Proof: Let (a_n) be a sequence of real numbers. We first try to find an increasing subsequence. If we just proceed naïvely, choosing a_{n_1} and seeking $n_2 > n_1$ satisfying $a_{n_2} \geq a_{n_1}$, we shall become unable to proceed if any of our suffices n_j has the property that $\forall n > n_j \ a_n < a_{n_j}$. This suggests that we look at the set of such points, so let $S = \{n \in \mathbf{N}: \forall m > n, a_m < a_n\}$, the set of suffices such that the corresponding term is greater than all subsequent terms.

If S is finite (including the case $S = \varnothing$) then choose n_1 to be greater than all the elements of S. Thus $n_1 \notin S$. By definition, then, $\exists n_2 > n_1$ s.t. $a_{n_2} \geq a_{n_1}$. Since $n_2 > n_1$, $n_2 \notin S$ and $\exists n_3 > n_2$ s.t. $a_{n_3} \geq a_{n_2}$. Proceeding in this way we obtain an increasing subsequence (a_{n_k}) of (a_n).

If S is infinite let $n_1 \in S$. For $n_k \in S$ let n_{k+1} be chosen to belong to S and satisfy $n_{k+1} > n_k$. This is possible for all k since S is infinite. This gives a subsequence (a_{n_k}) with the property that $\forall k \in \mathbf{N} \ a_{n_{k+1}} < a_{n_k}$ (since $n_k \in S$), a strictly decreasing sequence. $\square$

Lemma 11.4   A subsequence of a convergent sequence converges to the same limit as the whole sequence.

Proof: Let $x_n \to x$ as $n \to \infty$, where (x_n) is a sequence of real numbers. Let (x_{n_k}) be a subsequence of (x_n), so $\forall k \in \mathbf{N} \ n_k \geq k$.

Let $\varepsilon > 0$. $\exists N$ s.t. $\forall n \geq N \ |x_n - x| < \varepsilon$, whence, for this N,

$$k \geq N \Rightarrow n_k \geq k \geq N \Rightarrow |x_{n_k} - x| < \varepsilon.$$

Thus $x_{n_k} \to x$ as $k \to \infty$. $\square$

This result would be equally true of a sequence in $\mathbf{R}^m$.

Definition   A sequence $(\mathbf{x}_n)$ in $\mathbf{R}^m$ is said to be **bounded** if $(\|\mathbf{x}_n\|)$ is a bounded sequence in $\mathbf{R}$. A subset $A \subset \mathbf{R}^m$ is said to be **bounded** if $\{\|\mathbf{x}\| : \mathbf{x} \in A\}$ is a bounded subset of $\mathbf{R}$.

Theorem 11.5   (Bolzano–Weierstrass Theorem)   Every bounded sequence in $\mathbf{R}^m$ has a convergent subsequence.

Proof: We first take the simpler case of a sequence of real numbers. Let (a_n) be a bounded sequence of real numbers. Then there is a subsequence (a_{n_k}) which is either increasing or decreasing. Since (a_n) is bounded, so is (a_{n_k}), hence (a_{n_k}), being bounded and either increasing or decreasing, is convergent.

Now let $(\mathbf{x}_n)$ be a bounded sequence in $\mathbf{R}^m$; suppose that $\forall n \in \mathbf{N}$ $\|\mathbf{x}_n\| \leq K$. For each n, let $\mathbf{x}_n = (x_1^{(n)}, \ldots, x_m^{(n)})$. The sequence $(x_1^{(n)})$ (as n varies) is a sequence of real numbers, and since $\forall n$, $|x_1^{(n)}| \leq \|\mathbf{x}_n\| \leq K$, it is a bounded sequence. Choose a subsequence $(x_1^{n_1(k)})$ which is convergent, which is possible by the last paragraph. Now $|x_2^{n_1(k)}| \leq \|\mathbf{x}_{n_1(k)}\| \leq K$ whence $(x_2^{n_1(k)})$ is a bounded sequence of real numbers, so it has a convergent subsequence $(x_2^{n_2(k)})$. Since $(x_1^{n_2(k)})$ is a subsequence of the convergent sequence $(x_1^{n_1(k)})$, it also converges. Proceeding in this way we obtain after m steps a sequence $(n_m(k))_{k=1}^\infty$ of integers such that for $j = 1, \ldots, m$ each sequence $(x_j^{n_m(k)})$ converges, to x_j say. Since $\|\mathbf{x}^{n_m(k)} - \mathbf{x}\| \to 0$ as $k \to \infty$, where $\mathbf{x} = (x_1, \ldots, x_m)$, the result is proved. $\square$

Theorem 11.6   (Uniform Continuity)   Let A be a closed, bounded subset of $\mathbf{R}^m$ and $f : A \to \mathbf{R}$ be continuous. Then f is uniformly continuous, i.e. $\forall \varepsilon > 0$ $\exists \delta > 0$ s.t. $\forall \mathbf{x}, \mathbf{y} \in A$ $\|\mathbf{x} - \mathbf{y}\| < \delta \Rightarrow |f(\mathbf{x}) - f(\mathbf{y})| < \varepsilon$.

Proof: Suppose the result were false, so that $\exists \varepsilon > 0$ s.t. $\forall \delta > 0$ $\exists \mathbf{x}, \mathbf{y} \in A$ s.t. $\|\mathbf{x} - \mathbf{y}\| < \delta$ and $|f(\mathbf{x}) - f(\mathbf{y})| \geq \varepsilon$. Let $\varepsilon > 0$ have the stated property. For each $n \in \mathbf{N}$, since $1/n > 0$, $\exists \mathbf{x}_n, \mathbf{y}_n \in A$ s.t. $\|\mathbf{x}_n - \mathbf{y}_n\| < 1/n$ and $|f(\mathbf{x}_n) - f(\mathbf{y}_n)| \geq \varepsilon$. Since A is bounded, $(\mathbf{x}_n)$ has a convergent subsequence, $(\mathbf{x}_{n_k})$; let $\mathbf{x}_{n_k} \to \mathbf{x}$ as $k \to \infty$. Since A is closed, $\mathbf{x} \in A$. Also, since

$$\|\mathbf{y}_{n_k} - \mathbf{x}\| \leq \|\mathbf{y}_{n_k} - \mathbf{x}_{n_k}\| + \|\mathbf{x}_{n_k} - \mathbf{x}\| < 1/n_k + \|\mathbf{x}_{n_k} - \mathbf{x}\|,$$

$\mathbf{y}_{n_k} \to \mathbf{x}$ as $k \to \infty$. But f is continuous at $\mathbf{x}$, so $\exists \delta_1 > 0$ s.t. $\|\mathbf{y} - \mathbf{x}\| < \delta_1 \Rightarrow |f(\mathbf{y}) - f(\mathbf{x})| < \varepsilon/2$. Choose k so large that $\|\mathbf{x}_{n_k} - \mathbf{x}\| < \delta_1$ and $\|\mathbf{y}_{n_k} - \mathbf{x}\| < \delta_1$, then

$$|f(\mathbf{x}_{n_k}) - f(\mathbf{y}_{n_k})| \leq |f(\mathbf{x}_{n_k}) - f(\mathbf{x})| + |f(\mathbf{x}) - f(\mathbf{y}_{n_k})| < \varepsilon/2 + \varepsilon/2.$$

This is a contradiction, since by definition $|f(\mathbf{x}_{n_k}) - f(\mathbf{y}_{n_k})| \geq \varepsilon$. $\square$

In common with the corresponding result for functions of one variable, Theorem 10.6, the significant result here is that the δ above can be chosen to be independent of $\mathbf{x}$ and $\mathbf{y}$. Let $A = [a, b] \times [c, d]$ and $f : A \to \mathbf{R}$ be continuous. A is closed and bounded; to see the closure, notice that $p_i : \mathbf{R}^2 \to \mathbf{R}$ given by $p_i(x_1, x_2) = x_i$ is continuous and $A = p_1^{-1}([a, b]) \cap p_2^{-1}([c, d])$. Therefore the Theorem applies and f is uniformly continuous. Let $\varepsilon > 0$.

Then we can chose $\delta > 0$ such that $\forall x, y \in A$ $\|x - y\| < \delta \Rightarrow |f(x) - f(y)| < \varepsilon$. In particular, if $|x_1 - x_2| < \delta/2$ and $|y_1 - y_2| < \delta/2$, so that $\|(x_1, y_1) - (x_2, y_2)\| < \delta$, then $|f(x_1, y_1) - f(x_2, y_2)| < \varepsilon$. It is worth noticing that if $|x_1 - x_2| < \delta$ then, for all $y \in [c, d]$, $|f(x_1, y) - f(x_2, y)| < \varepsilon$, so that we can use the same value of δ for all $y \in [c, d]$ to ensure that $|f(x_1, y) - f(x_2, y)| < \varepsilon$. One use of this is:

Theorem 11.7   Let $f : [a, b] \times [c, d] \to \mathbf{R}$ be continuous and define $F : [a, b] \to \mathbf{R}$ by

$$F(x) = \int_c^d f(x, y)\, dy.$$

F is continuous.

Proof: Notice first that, for each $x \in [a, b]$, $y \mapsto f(x, y)$ is continuous, hence integrable, so the expression for F is meaningful.

Let $\varepsilon > 0$. By the uniform continuity of f on $[a, b] \times [c, d]$, $\exists \delta > 0$ s.t. $\forall x, x' \in [a, b] \times [c, d]$ $\|x - x'\| < \delta \Rightarrow |f(x) - f(x')| < \varepsilon/(d - c)$. Therefore let $x, x' \in [a, b]$ with $|x - x'| < \delta$. Then $\forall y \in [c, d]$ $|f(x, y) - f(x', y)| < \varepsilon/(d - c)$, so that $|F(x) - F(x')| \leq \int_c^d |f(x, y) - f(x', y)|\, dy < \varepsilon$.

Since $\varepsilon > 0$ was arbitrary we have shown that $\forall \varepsilon > 0 \exists \delta > 0$ s.t. $\forall x, x' \in [a, b]$ $|x - x'| < \delta \Rightarrow |F(x) - F(x')| < \varepsilon$ so F is (uniformly) continuous. $\square$

Obviously, the analogous result, obtaining a function of y by integrating with respect to x, is true. More generally, if f is a continuous function of the m variables $x_1, \ldots, x_m$ in $[a_1, b_1] \times \cdots \times [a_m, b_m]$, the function obtained by integrating with respect to one of these is continuous with respect to the remaining variables.

That a continuous function on a closed, bounded set is bounded is now easy:

Theorem 11.8   Let A be a closed, bounded subset of $\mathbf{R}^m$ and $f : A \to \mathbf{R}$ be continuous. Then f is a bounded function and f attains its supremum and infimum at points of A.

Proof: By Theorem 11.6, f is uniformly continuous, so $\exists \delta > 0$ s.t. $\forall x, y \in A$, $\|x - y\| < \delta \Rightarrow |f(x) - f(y)| < 1 \Rightarrow |f(x)| < |f(y)| + 1$.

Since A is bounded, there is an M for which $x \in A \Rightarrow \|x\| \leq M$ so $x = (x_1, \ldots, x_m) \in A \Rightarrow |x_i| \leq M$ $i = 1, 2, \ldots, m$. Thus A is a subset of the 'cube' $[-M, M] \times [-M, M] \times \cdots \times [-M, M]$. Let $a_0 = -M < a_1 < \cdots < a_r = M$ be such that, for each i, $a_i - a_{i-1} < \delta/m$, so if $x \in [-M, M]$, $x \in [a_{i-1}, a_i]$ for some i. Therefore

$$A = \bigcup_{i=1}^s A_i$$

where each A_i is the intersection of A with a set of the form $[a_{i_1-1}, a_{i_1}] \times \cdots \times [a_{i_m-1}, a_{i_m}]$ and we omit the set from the list if it is empty. (See Fig. 11.1.)

Choose $z_i = (z_1^{(i)}, \ldots, z_m^{(i)}) \in A_i$ for each i. If $x = (x_1, \ldots, x_m) \in A_i$, then, $|z_j^{(i)} - x_j| < \delta/m$ for each j, $\|x - z_i\| < \delta$ and so $|f(x)| < |f(z_i)| + 1$. Thus

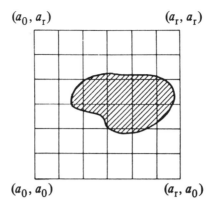

(a_0, a_r) (a_r, a_r)

(a_0, a_0) (a_r, a_0)

Fig. 11.1

$\forall \mathbf{x} \in A, |f(\mathbf{x})| < \max_{1 \le i \le s} (|f(\mathbf{z}_i)| + 1)$, since $\mathbf{x}$ must belong to some A_i. It follows that f is bounded.

To show the supremum is attained, let $a = \sup\{f(\mathbf{x}) : \mathbf{x} \in A\}$. Then if a is not attained and we set $g(\mathbf{x}) = 1/(a - f(\mathbf{x}))$, g is continuous on A and thus bounded. This gives a contradiction in exactly the same way as in the case of a function of one variable. □

As with continuity, we may consider the differentiation of a function of several variables either by fixing all but one of them and differentiating the resulting function of one variable or by considering a differentiation process where all the variables are treated simultaneously. As with continuity again, these two processes differ.

Definitions   Suppose that A is a subset of $\mathbf{R}^m$, $f : A \to \mathbf{R}$ and that $\mathbf{a} \in A$. If the function g_i defined by $g_i(x) = f(a_1, \dots, a_{i-1}, x, a_{i+1}, \dots, a_m)$ is differentiable at a_i, then we say the **partial derivative** $\partial f/\partial x_i$ exists at $\mathbf{a} = (a_1, \dots, a_m)$ and its value is $g_i'(a_i)$. f is said to be **differentiable** at $\mathbf{a}$ if for some $h > 0$ $\{\mathbf{x} \in \mathbf{R}^m : \|\mathbf{x} - \mathbf{a}\| < h\} \subset A$ and, for some constants $\beta_1, \dots, \beta_m$,

$$f(\mathbf{x}) = f(\mathbf{a}) + \sum_{i=1}^{m} \beta_i(x_i - a_i) + \theta(\mathbf{x})$$

where $\theta(\mathbf{x})/\|\mathbf{x} - \mathbf{a}\| \to 0$ as $\mathbf{x} \to \mathbf{a}$.

Partial derivatives are the type of derivative we have met before, of functions of one variable, and we can apply our existing theorems to them. Notice that there is no need for all partial derivatives to exist: if $f(x_1, x_2) = x_1 + |x_2|$ then $\partial f/\partial x_1$ exists at $(0, 0)$ but $\partial f/\partial x_2$ does not.

Differentiability of f as a whole is about the approximation of $f(\mathbf{x}) - f(\mathbf{a})$ by a linear function of $\mathbf{x} - \mathbf{a}$, the error θ being 'small' in the sense stated. This is often expressed in terms of 'differentials' as $df = \sum_{i=1}^{m} \beta_i(dx_i)$ where, as we shall see, β_i turns out to equal $\partial f/\partial x_i$ at $\mathbf{a}$. The more precise statement is the one we have made about approximating $f(\mathbf{x}) - f(\mathbf{a})$ by the linear function $\sum_{i=1}^{m} \beta_i(x_i - a_i)$.

Example Let $f(x_1, x_2) = x_1 x_2$. It is easy to see that at the point (a_1, a_2), $\partial f/\partial x_1 = a_2$ and $\partial f/\partial x_2 = a_1$, so f possesses partial derivatives.

$$f(x_1, x_2) - f(a_1, a_2) = x_1 x_2 - a_1 a_2 = (x_1 - a_1)a_2 + (x_2 - a_2)a_1 + \theta(x_1, x_2)$$

where $\theta(x_1, x_2) = (x_1 - a_1)(x_2 - a_2)$. At this point we recall that if $u, v \in \mathbf{R}$ $|2uv| \leq u^2 + v^2$ so $|\theta(x_1, x_2)|/\|\mathbf{x} - \mathbf{a}\| \leq \frac{1}{2}\sqrt{((x_1 - a_1)^2 + (x_2 - a_2)^2)}$ and $\theta(x_1, x_2)/\|\mathbf{x} - \mathbf{a}\| \to 0$ as $\mathbf{x} \to \mathbf{a}$. f is differentiable at (a_1, a_2) and the constants β_1, β_2 equal $(\partial f/\partial x_1)(a_1, a_2)$ and $(\partial f/\partial x_2)(a_1, a_2)$ respectively.

(In practice, writing $x_1 - a_1 = r\cos\theta$, $x_2 - a_2 = r\sin\theta$ is illuminating in the above calculation.)

Lemma 11.9 Let A be a subset of $\mathbf{R}^m$ and $\{\mathbf{x} : \|\mathbf{x} - \mathbf{a}\| < h\} \subset A$, where $h > 0$. If $f : A \to \mathbf{R}$ is differentiable at $\mathbf{a}$ then all the partial derivatives at $\mathbf{a}$ exist and $(\partial f/\partial x_i)(a_1, \ldots, a_m) = \beta_i$ $(i = 1, 2, \ldots, m)$ where β_i is the number in the definition of differentiability.

Conversely, if all the partial derivatives of f exist at all points of A, and are continuous at $\mathbf{a}$, then f is differentiable at $\mathbf{a}$.

Proof: Suppose that f is differentiable at $\mathbf{a}$ and that

$$f(\mathbf{x}) = f(\mathbf{a}) + \sum_{i=1}^{m} \beta_i(x_i - a_i) + \theta(\mathbf{x})$$

where $\theta(\mathbf{x})/\|\mathbf{x} - \mathbf{a}\| \to 0$ as $\mathbf{x} \to \mathbf{a}$. Then let $\mathbf{x} = (a_1, \ldots, a_{i-1}, t, a_{i+1}, \ldots, a_m)$ and $g_i(t) = f(a_1, \ldots, a_{i-1}, t, a_{i+1}, \ldots, a_m)$. Since $\|\mathbf{x} - \mathbf{a}\| = |t - a_i|$ we have

$$g_i(t) = g_i(a_i) + \beta_i(t - a_i) + \theta_i(t)$$

where $\theta_i(t) = \theta(\mathbf{x})$ and $\theta_i(t)/|t - a_i| \to 0$ as $t \to a_i$. From this we see that g_i is differentiable at a_i, and that $\beta_i = g_i'(a_i) = (\partial f/\partial x_i)(a_1, \ldots, a_m)$.

Now suppose that all the partial derivatives exist, and that they are continuous at $\mathbf{a}$. To deduce information about $f(\mathbf{x}) - f(\mathbf{a})$ we need to re-express this in a form where in each part only one variable changes, so if $\|\mathbf{x} - \mathbf{a}\| < h$

$$f(x_1, \ldots, x_m) - f(a_1, \ldots, a_m) = \sum_{i=1}^{m} (f(x_1, \ldots, x_i, a_{i+1}, \ldots, a_m)$$

$$- f(x_1, \ldots, x_{i-1}, a_i, \ldots, a_m))$$

$$= \sum_{i=1}^{m} \frac{\partial f}{\partial x_i}(x_1, \ldots, x_{i-1}, \xi_i, a_{i+1}, \ldots, a_m)$$

$$\cdot (x_i - a_i), \tag{1}$$

obtained by applying the one-variable Mean Value Theorem to the function $t \mapsto f(x_1, \ldots, x_{i-1}, t, a_{i+1}, \ldots, a_m)$; ξ_i will be between a_i and x_i (and will depend on $x_1, \ldots, x_i$). In this notation, we have, for $\|\mathbf{x} - \mathbf{a}\| < h$, that $f(\mathbf{x}) = f(\mathbf{a}) + \sum_{i=1}^{m} (\partial f/\partial x_i)(\mathbf{a}) \cdot (x_i - a_i) + \theta(\mathbf{x})$ where

$$\theta(\mathbf{x}) = \sum_{i=1}^{m} \left(\frac{\partial f}{\partial x_i}(x_1, \ldots, x_{i-1}, \xi_i, a_{i+1}, \ldots, a_m) - \frac{\partial f}{\partial x_i}(a_1, \ldots, a_m) \right)(x_i - a_i).$$

Let $\varepsilon > 0$. Then, by continuity of $\partial f / \partial x_i$ at $\mathbf{a}$, $\exists \delta > 0$ s.t. $\| \mathbf{y} - \mathbf{a} \| < \delta \Rightarrow$ for all i $|(\partial f / \partial x_i)(\mathbf{y}) - (\partial f / \partial x_i)(\mathbf{a})| < \varepsilon / m$. Choose $\| \mathbf{x} - \mathbf{a} \| < \delta$, so, with ξ_i as in (1),

$$\| (x_1, \ldots, x_{i-1}, \xi_i, a_{i+1}, \ldots, a_m) - (a_1, \ldots, a_m) \| \leq \| \mathbf{x} - \mathbf{a} \| < \delta$$

and

$$|\theta(\mathbf{x})| \leq \sum_{i=1}^{m} \left| \frac{\partial f}{\partial x_i} (x_1, \ldots, x_{i-1}, \xi_i, a_{i+1}, \ldots, a_m) - \frac{\partial f}{\partial x_i} (a_1, \ldots, a_m) \right| \cdot |x_i - a_i|$$

$$< \varepsilon / m \sum_{i=1}^{m} |x_i - a_i| \leq \varepsilon \| \mathbf{x} - \mathbf{a} \| \quad (\text{since } |x_i - a_i| \leq \| \mathbf{x} - \mathbf{a} \|).$$

So $0 < \| \mathbf{x} - \mathbf{a} \| < \delta \Rightarrow |\theta(\mathbf{x})| / \| \mathbf{x} - \mathbf{a} \| < \varepsilon$, and, since $\varepsilon > 0$ was arbitrary, $\theta(x) / \| \mathbf{x} - \mathbf{a} \| \to 0$ as $\mathbf{x} \to \mathbf{a}$. $\square$

The condition for differentiability may be put in a version which is co-ordinate-free by observing that $\sum_{i=1}^{m} \beta_i (x_i - a_i)$ is the form of a typical linear function from $\mathbf{R}^m$ to $\mathbf{R}$ evaluated at $\mathbf{x} - \mathbf{a}$; we shall not pursue this. Notice that differentiability is a stronger condition than the existence of the partial derivatives. It is differentiability that we require in order to establish the chain rule for functions of several variables, but we have a few items to tidy up first.

Examples   Let $f(x, y) = xy / \sqrt{(x^2 + y^2)}$ $((x, y) \neq (0, 0))$ and $f(0, 0) = 0$. Then $(\partial f / \partial x)(0, 0) = \lim_{x \to 0} (f(x, 0) - f(0, 0)) / x = 0$ and $(\partial f / \partial y)(0, 0) = 0$. f is not, however, differentiable at $(0, 0)$, since $f(x, y) - f(0, 0) - \beta_1 x - \beta_2 y = xy / \sqrt{(x^2 + y^2)}$ where $\beta_1 = (\partial f / \partial x)(0, 0)$ and $\beta_2 = (\partial f / \partial y)(0, 0)$, and $xy / \sqrt{(x^2 + y^2)} \nrightarrow 0$ as $(x, y) \to (0, 0)$.

The existence of partial derivatives of f does not imply that f is continuous. If $\partial f / \partial x_1$ exists at a point $\mathbf{a}$, this means that, *as a function of x_1 only, f is differentiable* and so f is continuous with respect to the variable x_1. Therefore the existence of all the partial derivatives only guarantees that f is continuous with respect to each variable separately. Consideration of the function defined by $f(x, y) = xy / (x^2 + y^2)$ for $(x, y) \neq (0, 0)$, $f(0, 0) = 0$ gives a function which is not continuous at $(0, 0)$ but $(\partial f / \partial x)$ and $(\partial f / \partial y)$ exist at all points.

If f is differentiable at $\mathbf{a}$, then f is continuous at $\mathbf{a}$. By definition of differentiability, $f(\mathbf{x}) = f(\mathbf{a}) + \sum_{i=1}^{m} \beta_i (x_i - a_i) + \theta(\mathbf{x})$ where $\theta(\mathbf{x}) / \| \mathbf{x} - \mathbf{a} \| \to 0$ as $\mathbf{x} \to \mathbf{a}$. Since for each i, $|x_i - a_i| \leq \| \mathbf{x} - \mathbf{a} \|$, $\sum_{i=1}^{m} \beta_i (x_i - a_i) \to 0$ as $\mathbf{x} \to \mathbf{a}$, and also $\theta(\mathbf{x}) = (\theta(\mathbf{x}) / \| \mathbf{x} - \mathbf{a} \|) . \| \mathbf{x} - \mathbf{a} \| \to 0$, so f is continuous at $\mathbf{a}$.

Theorem 11.10   (The Chain Rule)   Suppose that $\mathbf{b} \in B \subset \mathbf{R}^m$ and that $f: B \to \mathbf{R}$ is differentiable at $\mathbf{b}$. Let $g_1, \ldots, g_m : A \to \mathbf{R}$ be m functions, differentiable at $\mathbf{a} \in A \subset \mathbf{R}^n$ and with $g_i(\mathbf{a}) = b_i$ ($i = 1, 2, \ldots, m$). Then the function $h: A \to \mathbf{R}$ where

$$h(x_1, \ldots, x_n) = f(g_1(x_1, \ldots, x_n), \ldots, g_m(x_1, \ldots, x_n))$$

is differentiable at $\mathbf{a}$ and

$$\frac{\partial h}{\partial x_j}(\mathbf{a}) = \sum_{i=1}^{m} \frac{\partial f}{\partial y_i}(\mathbf{g(a)}) \cdot \frac{\partial g_i}{\partial x_j}(\mathbf{a})$$

(where $g(a) = (g_1(a), \ldots, g_m(a))$). Here f is regarded as a function of $(y_1, \ldots, y_m)$.

Proof: By the assumption of differentiability, $f(\mathbf{y}) = f(\mathbf{b}) + \sum_{i=1}^m \beta_i(y_i - b_i) + \theta(\mathbf{y})$ and $g_i(\mathbf{x}) = g_i(\mathbf{a}) + \sum_{j=1}^n \alpha_{ij}(x_j - a_j) + \phi_i(\mathbf{x})$ where $\theta(\mathbf{y})/\|\mathbf{y} - \mathbf{b}\| \to 0$ as $\mathbf{y} \to \mathbf{b}$ and $\phi_i(\mathbf{x})/\|\mathbf{x} - \mathbf{a}\| \to 0$ as $\mathbf{x} \to \mathbf{a}$ $(i = 1, \ldots, m)$. Then

$$h(\mathbf{x}) = f(\mathbf{g}(\mathbf{x})) = f(\mathbf{g}(\mathbf{a})) + \sum_{i=1}^m \beta_i(g_i(\mathbf{x}) - g_i(\mathbf{a})) + \theta(\mathbf{g}(\mathbf{x}))$$

$$= f(\mathbf{g}(\mathbf{a})) + \sum_{i=1}^m \beta_i\left(\sum_{j=1}^n \alpha_{ij}(x_j - a_j)\right) + \sum_{i=1}^m \beta_i\phi_i(\mathbf{x}) + \theta(\mathbf{g}(\mathbf{x}))$$

$$= f(\mathbf{g}(\mathbf{a})) + \sum_{j=1}^n \gamma_j(x_j - a_j) + \psi(\mathbf{x}), \tag{1}$$

where $\gamma_j = \sum_{i=1}^m \beta_i\alpha_{ij}$ and $\psi(\mathbf{x}) = \sum_{i=1}^m \beta_i\phi_i(\mathbf{x}) + \theta(\mathbf{g}(\mathbf{x}))$. We have to show $\psi(\mathbf{x})/\|\mathbf{x} - \mathbf{a}\| \to 0$ as $\mathbf{x} \to \mathbf{a}$. From what we already know of ϕ_i, we need only prove $\theta(\mathbf{g}(\mathbf{x}))/\|\mathbf{x} - \mathbf{a}\| \to 0$. Let $M = (1 + \max_i \sum_j |\alpha_{ij}|)\sqrt{m}$. $\exists \delta_1 > 0$ s.t. $0 < \|\mathbf{y} - \mathbf{b}\| < \delta_1 \Rightarrow |\theta(\mathbf{y})|/\|\mathbf{y} - \mathbf{b}\| < \varepsilon/M$, since $\theta(\mathbf{y})/\|\mathbf{y} - \mathbf{b}\| \to 0$, so

$$\|\mathbf{y} - \mathbf{b}\| < \delta_1 \Rightarrow |\theta(\mathbf{y})| \le \varepsilon\|\mathbf{y} - \mathbf{b}\|/M. \tag{2}$$

Since $\phi_i(\mathbf{x})/\|\mathbf{x} - \mathbf{a}\| \to 0$ as $\mathbf{x} \to \mathbf{a}$,

$$\exists \delta_2 > 0 \text{ s.t. } 0 < \|\mathbf{x} - \mathbf{a}\| < \delta_2 \Rightarrow |\phi_i(\mathbf{x})|/\|\mathbf{x} - \mathbf{a}\| < 1 \quad (i = 1, 2, \ldots, m)$$

$$\Rightarrow |g_i(\mathbf{x}) - g_i(\mathbf{a})| \le \sum_{j=1}^n |\alpha_{ij}|\,|x_j - a_j| + |\phi_i(\mathbf{x})|$$

$$< \left(1 + \sum_{j=1}^n |\alpha_{ij}|\right)\|\mathbf{x} - \mathbf{a}\|$$

$$\Rightarrow \|\mathbf{g}(\mathbf{x}) - \mathbf{g}(\mathbf{a})\| \le M\|\mathbf{x} - \mathbf{a}\|. \tag{3}$$

Thus

$$0 < \|\mathbf{x} - \mathbf{a}\| < \min(\delta_2, \delta_1/M) \Rightarrow \|\mathbf{g}(\mathbf{x}) - \mathbf{g}(\mathbf{a})\| < \delta_1 \text{ by (3)}$$

$$\Rightarrow |\theta(\mathbf{g}(\mathbf{x}))| < \varepsilon\|\mathbf{g}(\mathbf{x}) - \mathbf{g}(\mathbf{a})\|/M \le \varepsilon\|\mathbf{x} - \mathbf{a}\|$$

$$\text{by (2) and (3).}$$

Therefore $\theta(\mathbf{g}(\mathbf{x}))/\|\mathbf{x} - \mathbf{a}\| \to 0$ as $\mathbf{x} \to \mathbf{a}$, so h is differentiable, by (1).

Finally, we see that γ_j which is $(\partial h/\partial x_j)(\mathbf{a})$, equals $\sum_{i=1}^m \beta_i\alpha_{ij}$, that is,

$$\sum_{i=1}^m \frac{\partial f}{\partial y_i}(\mathbf{g}(\mathbf{a}))\cdot\frac{\partial g_i}{\partial x_j}(\mathbf{a}). \quad \square$$

The Chain Rule for functions of several variables is easiest remembered in the form where f is a function of the variables $y_1, \ldots, y_m$ and each of $y_1, \ldots, y_m$ is a function of $x_1, \ldots, x_n$ giving, in the standard form,

$$\frac{\partial f}{\partial x_j} = \sum_{i=1}^m \frac{\partial f}{\partial y_i}\cdot\frac{\partial y_i}{\partial x_j},$$

the rate of change of f with x_j being the sum of the rate of change of f with each co-ordinate times the rate of change of that co-ordinate with x_j. We have not chosen to state the theorem in this way because the symbol f is used here with two distinct meanings, and the same is true of y_i. This is clearer if we take an example.

Let $f(y_1, y_2) = y_1 y_2$ and let $g_1(x_1, x_2) = x_1 + x_2$, $g_2(x_1, x_2) = x_1 - x_2$. Then let $h(x_1, x_2) = f(g_1(x_1, x_2), g_2(x_1, x_2)) = x_1^2 - x_2^2$. The first thing to notice is that h is not the same function as f (e.g. $h(1, 2) \neq f(1, 2)$). The confusion usually arises in applications where the value of f represents some physical quantity and h represents the same quantity as a function of different variables, so that by '$\partial f/\partial x_1$' we may really mean $\partial h/\partial x_1$.

The second issue with the traditional notation is that the y_1 in $\partial f/\partial y_1$ is spurious; $\partial f/\partial y_1$ is a function of two variables, being that function obtained from f by fixing the second variable and differentiating with respect to the first. Since these variables are 'dummy' variables the fact that we have to attach a name to them is unfortunate. $\partial f/\partial y_1$ above is the function whose value at (a, b) is b. The Newtonian notation of f_{y_1} for $\partial f/\partial y_1$ is no better in this respect, while the various attempts to introduce a more rational notation have not penetrated far into practice so we shall just have to make the best of the situation. The principal analytical advice is to be quite clear which functions are distinct and, where possible, to use different symbols to distinguish them.

Example   Let f be a function of the two variables (x, y) and define $F(x, y) = f(y, x)$. Show that $(\partial F/\partial x)(a, b) = (\partial f/\partial y)(b, a)$. (This is, of course, deliberately awkward.)

Solution: Let $g_1(x, y) = y$, $g_2(x, y) = x$ so that $F(x, y) = f(g_1(x, y), g_2(x, y))$ and

$$\frac{\partial F}{\partial x}(a, b) = \frac{\partial f}{\partial x}(\mathbf{g}(a, b)) \cdot \frac{\partial g_1}{\partial x}(a, b) + \frac{\partial f}{\partial y}(\mathbf{g}(a, b)) \cdot \frac{\partial g_2}{\partial x}(a, b)$$

$$= \frac{\partial f}{\partial x}(b, a) \cdot 0 + \frac{\partial f}{\partial y}(b, a) \cdot 1 = \frac{\partial f}{\partial y}(b, a).$$

(If it helps, you could write $\partial f/\partial g_1$ in place of $\partial f/\partial x$ and $\partial f/\partial g_2$ in place of $\partial f/\partial y$.)

Armed with the Chain Rule, it is now easy to apply earlier results.

Theorem 11.11 (Mean Value Theorem)   Let $f: A \to \mathbf{R}$ be continuous, $\mathbf{a}, \mathbf{b} \in A \subset \mathbf{R}^m$ and suppose that f is differentiable at all points on the straight line joining $\mathbf{a}$ to $\mathbf{b}$. Then, for some $\theta \in (0, 1)$

$$f(\mathbf{b}) - f(\mathbf{a}) = \sum_{i=1}^{m} \frac{\partial f}{\partial x_i}(\mathbf{a} + \theta(\mathbf{b} - \mathbf{a})) \cdot (b_i - a_i)$$

Proof: Define $F(t) = f(\mathbf{a} + t(\mathbf{b} - \mathbf{a}))$ for $t \in [0, 1]$. F is continuous on $[0, 1]$ and differentiable on $(0, 1)$ and $F'(t) = \sum_{i=1}^{m} (\partial f/\partial x_i)(\mathbf{a} + t(\mathbf{b} - \mathbf{a})) \cdot (b_i - a_i)$ by the Chain Rule. By the ordinary Mean Value Theorem $\exists \theta \in (0, 1)$ such that $F(1) - F(0) = F'(\theta)$. $\quad \square$

Definition   Let $A \subset \mathbf{R}^m$ and $f : A \to \mathbf{R}$. If f is differentiable then each of the partial derivatives $\partial f / \partial x_i$ is a function from A to $\mathbf{R}$ and it may in turn be differentiable. We define

$$\frac{\partial^2 f}{\partial x_j \partial x_i}$$

to be

$$\frac{\partial}{\partial x_j} \left(\frac{\partial f}{\partial x_i} \right),$$

the jth partial derivative of $\partial f / \partial x_i$, when this exists. f is said to be twice differentiable if f and each of its partial derivatives $\partial f / \partial x_i$ are differentiable. The analogous definitions apply to higher order derivatives.

The 'mixed' partial derivatives $(\partial^2 f / \partial x_1 \partial x_2)$ and $(\partial^2 f / \partial x_2 \partial x_1)$ are not necessarily equal, since the two limiting processes involved are taken in the opposite order in the two cases. Under very mild additional assumptions, however, they are equal, as we shall show shortly. We shall devote a little effort to some results, all of which are essentially about reversing the order of limiting processes.

Theorem 11.12 (Differentiation under the Integral Sign) Suppose that $f : [a, b] \times [c, d] \to \mathbf{R}$, and that $\partial f / \partial x$ exists and is continuous (with respect to both variables jointly). Then, with $F(x) = \int_c^d f(x, t) \, dt$, F is differentiable and $F'(x) = \int_c^d (\partial f / \partial x)(x, t) \, dt$.

Proof: We have to show that the derivative of the integral and the integral of the derivative are equal. Fix $x_0 \in (a, b)$; if x_0 is an endpoint, the usual minor changes are needed. Then, applying the Mean Value Theorem to $x \mapsto f(x, t)$ we see that, for each t, there is some ξ_t between x and x_0 satisfying $f(x, t) - f(x_0, t) = (x - x_0)(\partial f / \partial x)(\xi_t, t)$ and

$$F(x) - F(x_0) = \int_c^d (x - x_0) \frac{\partial f}{\partial x} (\xi_t, t) \, dt.$$

Let $\varepsilon > 0$. Then, by the uniform continuity of $\partial f / \partial x$ on $[a, b] \times [c, d]$, $\exists \delta > 0$ s.t. $\| (x, y) - (x', y') \| < \delta \Rightarrow |(\partial f / \partial x)(x, y) - (\partial f / \partial x)(x', y')| < \varepsilon / (d - c)$. Let $0 < |x - x_0| < \delta$. Then

$$\left| \frac{F(x) - F(x_0)}{x - x_0} - \int_c^d \frac{\partial f}{\partial x} (x_0, t) \, dt \right| \leq \int_c^d \left| \frac{\partial f}{\partial x} (\xi_t, t) - \frac{\partial f}{\partial x} (x_0, t) \right| dt < \varepsilon,$$

since for each t, ξ_t is between x_0 and x, so $\| (\xi_t, t) - (x_0, t) \| < |x - x_0| < \delta$, whence $\forall t \in [c, d]$ $|(\partial f / \partial x)(\xi_t, t) - (\partial f / \partial x)(x_0, t)| < \varepsilon / (d - c)$. This establishes that $F'(x_0) = \int_c^d (\partial f / \partial x)(x_0, t) \, dt$. $\quad \square$

Corollary   With f as above, if $\phi, \psi : [a, b] \to [c, d]$ are differentiable and

$F(x) = \int_{\phi(x)}^{\psi(x)} f(x, t) \, dt$ then

$$F'(x) = \int_{\phi(x)}^{\psi(x)} \frac{\partial f}{\partial x}(x, t) \, dt + f(x, \psi(x)) \cdot \psi'(x) - f(x, \phi(x)) \cdot \phi'(x).$$

Proof: Let $G(x, y, z) = \int_{y}^{z} f(x, t) \, dt$. $\partial G / \partial x$ exists by the Theorem, while $\partial G / \partial y = - f(x, y)$, $\partial G / \partial z = f(x, z)$ by the Fundamental Theorem of Calculus. All three partial derivatives are continuous, so we may apply the Chain Rule to $F(x) = G(x, \phi(x), \psi(x))$. □

Theorem 11.13   Let $f : [a, b] \times [c, d] \to \mathbf{R}$ be continuous. Then

$$\int_{a}^{b} \left\{ \int_{c}^{d} f(x, y) \, dy \right\} dx = \int_{c}^{d} \left\{ \int_{a}^{b} f(x, y) \, dx \right\} dy.$$

Proof: Define $g(x, z) = \int_{c}^{z} f(x, y) \, dy$. Then $(\partial g / \partial z) = f$, which is continuous. Now let $G(z) = \int_{a}^{b} g(x, z) \, dx$, so by differentiation under the integral sign, $G'(z) = \int_{a}^{b} f(x, z) \, dx$. Since G' is continuous (Theorem 11.7),

$$\int_{c}^{d} \left\{ \int_{a}^{b} f(x, z) \, dx \right\} dz = \int_{c}^{d} G' = G(d) - G(c) = G(d) = \int_{a}^{b} \left\{ \int_{c}^{d} f(x, y) \, dy \right\} dx.$$

□

The result about inverting the order of integration remains true under more general circumstances, but it may fail for improper integrals. To see this, consider inverting the order of integration in $\int_{1}^{\infty} (\int_{1}^{\infty} (x - y)(x + y)^{-3} \, dx) \, dy$.

Theorem 11.14   Suppose that f is defined in the neighbourhood of the point (a, b) in $\mathbf{R}^2$, that is, on the set $\{(x, y) \in \mathbf{R}^2 : \| (x, y) - (a, b) \| < h\}$ for some $h > 0$, that $\partial f / \partial x$ and $\partial f / \partial y$ are continuous on this set and that one of $\partial^2 f / \partial x \partial y$ and $\partial^2 f / \partial y \partial x$ exists and is (jointly) continuous on this set. Then both $\partial^2 f / \partial x \partial y$ and $\partial^2 f / \partial y \partial x$ exist and are equal on this set.

Proof: Suppose that $\partial^2 f / \partial x \partial y$ is continuous in the neighbourhood of (a, b). Then if (x, y) is in this neighbourhood of (a, b), regarding y as fixed, we see, by differentiating under the integral sign, that

$$\int_{b}^{y} \frac{\partial^2 f}{\partial x \partial y}(x, t) \, dt = \frac{\partial}{\partial x} \int_{b}^{y} \frac{\partial f}{\partial y}(x, t) \, dt = \frac{\partial}{\partial x} (f(x, y) - f(x, b))$$

$$= \frac{\partial f}{\partial x}(x, y) - \frac{\partial f}{\partial x}(x, b).$$

(This uses the continuity of $\partial f / \partial y$; see problem 17.) Since this is true of all y sufficiently close to b, and since, for fixed x, the left-hand side is differentiable w.r.t. y, the right-hand side is differentiable w.r.t. y and

$$\frac{\partial^2 f}{\partial x \partial y}(x, y) = \frac{\partial}{\partial y} \int_{b}^{y} \frac{\partial^2 f}{\partial x \partial y}(x, t) \, dt = \frac{\partial^2 f}{\partial y \partial x}(x, y) - 0. \quad □$$

There are several possible minor variations on Theorem 11.14, which may easily be made to apply to functions of more than two variables, and to higher order partial derivatives. For most purposes it is sufficient to know that if the two mixed derivatives are (jointly) continuous they are equal.

Theorem 11.15 (Taylor's Theorem)   Suppose that $\mathbf{a} \in \mathbf{R}^m$ and that $f:D \to \mathbf{R}$ where $D = \{\mathbf{x} \in \mathbf{R}^m : \|\mathbf{x} - \mathbf{a}\| < \delta\}$ for some $\delta > 0$. Then if f possesses all partial derivatives of orders up to and including n and if all these partial derivatives are continuous, we have, for $\|\mathbf{h}\| < \delta$,

$$f(\mathbf{a} + \mathbf{h}) = f(\mathbf{a}) + \sum_{i=1}^{m} h_i \frac{\partial f}{\partial x_{i_{n-1}}} (\mathbf{a}) + \cdots$$

$$+ \sum_{i_1, i_2, \ldots, i_{n-1} = 1}^{m} \frac{h_{i_1} \cdots h_{i_{n-1}}}{(n-1)!} \cdot \frac{\partial^{n-1} f}{\partial x_{i_1} \cdots \partial x_{i_{n-1}}} (\mathbf{a}) + R_n$$

where, for some $\theta \in (0, 1)$,

$$R_n = \sum_{i_1, i_2, \ldots, i_n = 1}^{m} \frac{h_{i_1} \cdots h_{i_n}}{n!} \cdot \frac{\partial^n f}{\partial x_{i_1} \cdots \partial x_{i_n}} (\mathbf{a} + \theta \mathbf{h}).$$

Proof: Because all the various partial derivatives are continuous, all up to order $n - 1$ are differentiable as functions of m variables, so the Chain Rule, applied n times, shows that if $F(t) = f(\mathbf{a} + t\mathbf{h})$ $(0 \le t \le 1)$ then F is n-times differentiable. The Chain Rule shows $F'(t) = \sum_{i=1}^{m} h_i(\partial f/\partial x_i)$, $F''(t) = \sum_{i_1, i_2 = 1}^{m} h_{i_1} h_{i_2}(\partial^2 f/\partial x_{i_1} \partial x_{i_2})$, etc. The rest is messy but routine.   □

Example   Suppose that $f:\mathbf{R}^2 \to \mathbf{R}$ is differentiable. If f has a local maximum at (a, b) then the functions $x \mapsto f(x, b)$ and $y \mapsto f(a, y)$ have local maxima at $x = a$ and $y = b$ respectively so $\partial f/\partial x$ and $\partial f/\partial y$ are both zero at (a, b). This, of course, is true if f has a local minimum, so a necessary condition for a local maximum or minimum is that both (or all if there are more than two variables) the first order partial derivatives should be zero at that point.

Now suppose $(\partial f/\partial x)(a, b) = (\partial f/\partial y)(a, b) = 0$. To find sufficient conditions for a maximum, minimum etc., we presume that f has partial derivatives of order two and that these are continuous. Then, for $\|\mathbf{h}\|$ small enough,

$$f(a + h_1, b + h_2) = f(a, b) + \left(h_1^2 \frac{\partial^2 f}{\partial x^2}(\xi) + 2h_1 h_2 \frac{\partial^2 f}{\partial x \partial y}(\xi) + h_2^2 \frac{\partial^2 f}{\partial y^2}(\xi) \right) \Big/ 2$$

for some point $\xi = (a + \theta h_1, b + \theta h_2)$ with $0 < \theta < 1$. If we choose $\|\mathbf{h}\|$ sufficiently small, the term on the right will be close to

$$f(a, b) + (\alpha h_1^2 + 2\beta h_1 h_2 + \gamma h_2^2)/2$$

where $\alpha = \partial^2 f/\partial x^2$, $\beta = \partial^2 f/\partial x \partial y$ and $\gamma = \partial^2 f/\partial y^2$, all evaluated at (a, b), so we investigate expressions of the form $\alpha h_1^2 + 2\beta h_1 h_2 + \gamma h_2^2$. Such expressions are called quadratic forms and their algebraic theory is elegant and complete; we shall merely poach some fragments from it. Notice that $\alpha h_1^2 + 2\beta h_1 h_2 + \gamma h_2^2 = \alpha(h_1 + (\beta/\alpha)h_2)^2 + ((\alpha\gamma - \beta^2)/\alpha)h_2^2$ so that

(i) if $\alpha\gamma - \beta^2 > 0$ and $\alpha > 0$ then $\alpha h_1^2 + 2\beta h_1 h_2 + \gamma h_2^2 > 0$ unless $h_1 = h_2 = 0$, whence $f(a, b) + (\alpha h_1^2 + 2\beta h_1 h_2 + \gamma h_2^2)/2$ has a minimum at $h_1 = h_2 = 0$.

(ii) if $\alpha\gamma - \beta^2 > 0$ and $\alpha < 0$ then $\alpha h_1^2 + 2\beta h_1 h_2 + \gamma h_2^2 < 0$ unless $h_1 = h_2 = 0$ so $f(a, b) + (\alpha h_1^2 + 2\beta h_1 h_2 + \gamma h_2^2)/2$ has a maximum at $h_1 = h_2 = 0$.

(iii) if $\alpha\gamma - \beta^2 < 0$ then $f(a, b) + (\alpha h_1^2 + 2\beta h_1 h_2 + \gamma h_2^2)/2$ has neither a maximum nor a minimum. If $\alpha \neq 0$ this may be seen by considering the cases $(h_1, h_2) = (\lambda, 0)$ and $(-\lambda\beta/\alpha, \lambda)$, while re-expressing by taking a $\gamma(h_2 + (\beta/\gamma)h_1)^2$ term shows the same situation if $\gamma \neq 0$. The case $\alpha = \gamma = 0$, $\beta \neq 0$ is easy. Since $\alpha\gamma - \beta^2 > 0$ implies $\alpha \neq 0$, we have considered all cases where $\alpha\gamma - \beta^2 \neq 0$. In case (iii), f is said to have a saddle point.

These are the cases we can deal with, since the expression for $f(a + h_1, b + h_2)$ involves $\partial^2 f/\partial x^2$ etc. evaluated not at (a, b) but at $(a + \theta h_1, b + \theta h_2)$. If $\partial^2 f/\partial x^2$ is continuous and $\alpha = (\partial^2 f/\partial x^2)(a, b) \neq 0$, then for $\|\mathbf{h}\|$ small enough, $(\partial^2 f/\partial x^2)(a + \theta h_1, b + \theta h_2)$ will have the same sign as α; similarly, if $\alpha\gamma - \beta^2 \neq 0$, $(\partial^2 f/\partial x^2)(\partial^2 f/\partial y^2) - (\partial^2 f/\partial x\partial y)^2$ will have constant sign near (a, b). Since our arguments about the sign of $\alpha h_1^2 + 2\beta h_1 h_2 + \gamma h_2^2$ depend only on the sign of α and $\alpha\gamma - \beta^2$, the conclusions about the sign of $f(a + h_1, b + h_2) - f(a, b)$ are valid.

For the remaining cases, $\alpha\gamma - \beta^2 = 0$, and we see, for example, that $\alpha h_1^2 + 2\beta h_1 h_2 + \gamma h_2^2$ is always non-negative if $\alpha > 0$, but that there are non-zero values of h_1, h_2 for which the expression is zero. For these values of h_1, h_2, if we substitute the appropriate derivatives evaluated at $(a + \theta h_1, a + \theta h_2)$ we cannot be sure of the sign of $f(a + h_1, b + h_2) - f(a, b)$.

The results for a function of three or more variables are similar but a little more complicated.

At this point the techniques we have used are becoming somewhat strained, largely because in handling functions of several variables there is a degree of technicality and complication which can obscure matters. This is particularly true if one tries to consider whether a function whose domain is a subset of $\mathbf{R}^2$ and whose image is in $\mathbf{R}^2$ has an inverse function, where we need to treat simultaneously the two real-valued functions corresponding to the first and second co-ordinates of the image. This sort of issue is best tackled using a rather more abstract and sophisticated viewpoint which eliminates some of the technicalities, so we shall not pursue it here. A good source is Apostol (1974).

As an example of what can be done with our techniques, let us consider the existence and uniqueness of solutions of differential equations. The existence result is mainly of interest in awkward cases where solutions cannot easily be calculated but the uniqueness part is essential even when solutions can be found exactly, since it is the basis of our knowledge of when we have found all possible solutions.

Example   Let $f: [a, b] \times \mathbf{R} \to \mathbf{R}$ be continuous and satisfy the condition

$$\forall x \in [a, b] \quad \text{and} \quad \forall y_1, y_2 \in \mathbf{R} \quad |f(x, y_1) - f(x, y_2)| \leq M|y_1 - y_2|.$$

Then the differential equation

$$\frac{dy}{dx} = f(x, y(x)) \quad (a \le x \le b)$$

$$y(a) = \alpha$$

has a unique solution.

Remarks: The condition demanded of f may seem a little odd at first glance. If, however, f is differentiable and $\partial f/\partial y$ is bounded, the one-variable version of the Mean Value Theorem will show the truth of the condition assumed if $M = \sup\{|\partial f/\partial y(x, y)| : a \le x \le b, y \in \mathbf{R}\}$. The boundedness of $\partial f/\partial y$ would be automatic if $\partial f/\partial y$ were continuous and restricted to the domain $[a, b] \times [c, d]$ for some $[c, d]$, but we would then require that the values of all the functions y_n defined below lay in $[c, d]$, so the choice of a suitable interval is not entirely obvious. These issues can be successfully tackled in a study of differential equations but, since our purpose is to illustrate the analysis, we shall evade them by making the assumption stated.

Solution: Firstly notice that if y satisfies the differential equation then it is a continuous function satisfying

$$y(x) = \alpha + \int_a^x f(t, y(t))\,dt \quad (a \le x \le b). \tag{1}$$

Conversely, if y is a continuous function satisfying (1), then $y(a) = \alpha$ and, since the RHS is differentiable by the Fundamental Theorem of Calculus so is y, and y satisfies the differential equation. It is therefore enough to show that (1) has a unique continuous solution.

To simplify matters we shall initially assume $b - a < 1/M$.
Define

$$y_0(x) = \alpha, \quad y_{n+1}(x) = \alpha + \int_a^x f(t, y_n(t))\,dt. \tag{2}$$

Then, by induction, all the functions y_n are continuous. Let $d_n = \sup\{|y_n(t) - y_{n-1}(t)| : a \le t \le b\}$, so, since for $x \in [a, b]$,

$$|y_{n+1}(x) - y_n(x)| = \left|\int_a^x f(t, y_n(t)) - f(t, y_{n-1}(t))\,dt\right| \quad \text{by (2)}$$

$$\le \int_a^x |f(t, y_n(t)) - f(t, y_{n-1}(t))|\,dt$$

$$\le \int_a^x M|y_n(t) - y_{n-1}(t)|\,dt \le (b-a)Md_n,$$

we see that $d_{n+1} \le (b-a)Md_n$. On our assumption that $(b-a)M < 1$ it follows that $\sum d_n$ converges, and thus, by comparison, $\sum_n (y_n(x) - y_{n-1}(x))$ converges. But $\sum_{n=1}^N (y_n(x) - y_{n-1}(x)) = y_N(x) - y_0(x)$, so the convergence of this implies that, for each $x \in [a, b]$, $y_n(x)$ tends to a limit as $n \to \infty$. Let $y(x) = \lim_{n \to \infty} y_n(x)$.

We need to show y is continuous. Choose $x_0 \in [a, b]$ and let $\varepsilon > 0$. For all $x \in [a, b]$, $|y(x) - y_N(x)| = |\sum_{n=N+1}^{\infty} (y_n(x) - y_{n-1}(x))| \leq \sum_{n=N+1}^{\infty} d_n$, so choose N so that $\sum_{n=N+1}^{\infty} d_n < \varepsilon/3$. By the continuity of y_N $\exists \delta > 0$ s.t. $|x - x_0| < \delta \Rightarrow |y_N(x) - y_N(x_0)| < \varepsilon/3$. Then

$$|x - x_0| < \delta \Rightarrow |y(x) - y(x_0)| \leq |y(x) - y_N(x)| + |y_N(x) - y_N(x_0)|$$
$$+ |y_N(x_0) - y(x_0)|$$

$$< 2 \sum_{n=N+1}^{\infty} d_n + \varepsilon/3 < \varepsilon.$$

Since $\varepsilon > 0$ was arbitrary we have shown y continuous at x_0, and since x_0 was typical, y is continuous.

The function y is continuous so $\int_a^x f(t, y(t)) dt$ exists and we might hope this is the limit as $n \to \infty$ of $\int_a^x f(t, y_n(t)) dt$. As this conclusion is equivalent to interchanging the limiting operations of integrating and letting $n \to \infty$, we need to prove its truth. Choose N such that $\sum_{n=N+1}^{\infty} d_n < \varepsilon$, so that, as above,

$$\forall t \in [a, b] \quad \text{and} \quad \forall m \geq N \quad |y(t) - y_m(t)| \leq \sum_{n=m+1}^{\infty} |y_n(t) - y_{n-1}(t)|$$

$$\leq \sum_{n=m+1}^{\infty} d_n < \varepsilon.$$

Then, if $x \in [a, b]$ and $m \geq N$

$$\left| y_{m+1}(x) - \alpha - \int_a^x f(t, y(t)) dt \right| = \left| \int_a^x f(t, y_m(t)) - f(t, y(t)) dt \right|$$

$$\leq \int_a^x M |y_m(t) - y(t)| dt < M(b - a)\varepsilon.$$

Since $\varepsilon > 0$ was arbitrary, $y_{m+1}(x) \to \alpha + \int_a^x f(t, y(t)) dt$ as $m \to \infty$, whence $y(x) = \alpha + \int_a^x f(t, y(t)) dt$. As this is true for all $x \in [a, b]$, y satisfies (1), so a solution exists.

To show the solution is unique, let y and z be continuous solutions of (1), and let $d = \sup \{|y(t) - z(t)| : a \leq t \leq b\}$. Then if $x \in [a, b]$

$$|y(x) - z(x)| \leq \int_a^x |f(t, y(t)) - f(t, z(t))| dt \leq M(b - a)d.$$

Since this is true for all $x \in [a, b]$, $d \leq M(b - a)d$ which is impossible unless $d = 0$ (since $M(b - a) < 1$). Thus $d = 0$ and y and z are identical, proving the uniqueness.

Finally, suppose $b - a$ is no longer presumed less than $1/M$. Choose h to satisfy $0 < h < 1/M$ and apply the result on the interval $[a, a + h]$, obtaining a unique solution of the differential equation on $[a, a + h]$. Let $\alpha_1 = y(a + h)$ for the function just found and apply the result to the differential equation problem

$$\frac{dy}{dx} = f(x, y(x)) \quad (a + h \leq x \leq a + 2h), \quad y(a + h) = \alpha_1.$$

This has a unique solution y defined on $[a+h, a+2h]$. By proceeding in this way we obtain a solution y defined on $[a, b]$ and it turns out to be unique, though a little thought is needed to confirm the uniqueness overall from the uniqueness on the subintervals.

This result is fairly general in that the techniques may be used to prove results about second or higher order differential equations. The ideas here can be made more systematic and streamlined to minimise the amount of hard work involved, but for this the reader will have to await a more advanced analysis course. (See, for example, Taylor (1958) or Brown and Page (1970).)

Problems

1. In each case below, define $f(x, y)$ to have the value stated for $(x, y) \neq (0, 0)$ and to have the value 0 at $(0, 0)$. Determine whether f is continuous at $(0, 0)$:

$$x^2 y/(x^2 + y^2), \quad x/\sqrt{(x^2 + y^2)}, \quad (x + y)^2/\sqrt{(x^2 + y^2)}, \quad (x^2 - y^4)/(x^2 + y^4).$$

2. Let $f, g_1, g_2 : \mathbf{R}^2 \to \mathbf{R}$ be three continuous functions and define F by $F(x, y) = f(g_1(x, y), g_2(x, y))$. Show F is continuous. Prove also that if $h : \mathbf{R} \to \mathbf{R}$ is continuous so is H, where $H(x, y) = h(f(x, y))$.

3. Let f and g be defined as follows:

$$f(x, y) = \begin{cases} \sin(x + y)/(x + y) & (x + y \neq 0), \\ 1 & (x + y = 0), \end{cases}$$

$$g(x, y) = \begin{cases} \dfrac{\sin x - \sin y}{x - y} & (x \neq y) \\ \cos x & (x = y). \end{cases}$$

Show that both functions are continuous at $(0, 0)$.

4. Let $f : \mathbf{R}^m \to \mathbf{R}$ be continuous at $\mathbf{a}$. For $\boldsymbol{\theta} \in \mathbf{R}^m$, define $F : \mathbf{R} \to \mathbf{R}$ by $F(t) = F(\mathbf{a} + t\boldsymbol{\theta})$ and show that F is continuous at 0. The converse result is false; to see this define $f : \mathbf{R}^2 \to \mathbf{R}$ by

$$f(x, y) = \begin{cases} 1 & x^2 < y < 2x^2, \\ 0 & \text{otherwise.} \end{cases}$$ f is not continuous at $(0, 0)$ but

$\lim_{t \to 0} f(t\boldsymbol{\theta}) = 0$ for each vector $\boldsymbol{\theta} \in \mathbf{R}^2$.

5. Which of the following subsets of $\mathbf{R}^2$ are closed? $\{x \in \mathbf{R}^2 : \|x\| \leq 1\}$, $\{(x_1, x_2) : x_1 \geq 0\}$, $\{(x_1, x_2) : x_1^2 + x_2^2 < 1 \text{ or } (x_1^2 + x_2^2 = 1 \text{ and } x_1 \geq 0)\}$.

6. Suppose that (x_n) is a sequence of real numbers with the property that $\forall \varepsilon > 0 \, \exists N \text{ s.t. } \forall m, n \geq N \, |x_m - x_n| < \varepsilon$; such a sequence is called a Cauchy sequence. Prove that if (x_n) has a convergent subsequence, then (x_n) is itself convergent. By showing that (x_n) is a bounded sequence, deduce that it converges. (Thus every Cauchy sequence in $\mathbf{R}$ converges.)

7. Let $A = \{x \in \mathbf{R}^m : \|x - a\| \leq r\}$ and suppose that $f : A \to \mathbf{R}$ is continuous. If

$\mathbf{b}, \mathbf{c} \in A$ and $f(\mathbf{b}) < \gamma < f(\mathbf{c})$, prove that there is a point $\mathbf{x} \in A$ for which $f(\mathbf{x}) = \gamma$. (Hint: Define $F:[0, 1] \to \mathbf{R}$ s.t. $F(0) = f(\mathbf{b})$ and $F(1) = f(\mathbf{c})$?)

8. Calculate the partial derivatives $\partial f / \partial x$ and $\partial f / \partial y$ of the following functions at $(0, 0)$ and decide whether f is differentiable at $(0, 0)$. In all cases, $f(0, 0) = 0$ and for other values of (x, y), $f(x, y)$ is given by the formula stated: $x^2 y/(x^2 + y^2)$, $xy/\sqrt{(x^2 + y^2)}$, $x\sqrt{(x^2 + y^2)}$, $x^2 y^2/(x^2 + y^2)$.

9. Let $f:\mathbf{R}^m \to \mathbf{R}$ and $\mathbf{a} \in \mathbf{R}^m$. If $\boldsymbol{\theta} \in \mathbf{R}^m$ and $\|\boldsymbol{\theta}\| = 1$ we say that the directional derivative of f at $\mathbf{a}$ in the direction $\boldsymbol{\theta}$ is $F'(0)$ where $F(t) = f(\mathbf{a} + t\boldsymbol{\theta})$, when this exists. (So $\partial f/\partial x_1$ is the directional derivative in the direction $(1, 0, \ldots, 0)$.) Show that if f is differentiable, the directional derivative in the direction $\boldsymbol{\theta}$ is $\sum_{i=1}^m \theta_i \cdot (\partial f/\partial x_i)(\mathbf{a})$.

10. Let $f(x, y) = xy(x^2 - y^2)/(x^2 + y^2)$ if $(x, y) \neq (0, 0)$ and $f(0, 0) = 0$. Calculate the first and second order partial derivatives and show that $\partial^2 f/\partial x \partial y$ and $\partial^2 f/\partial y \partial x$ are unequal at $(0, 0)$.

11. Define $g_1, g_2:\mathbf{R}^2 \to \mathbf{R}$ by $g_1(x_1, x_2) = x_2$, $g_2(x_1, x_2) = x_1$, and set $G(x_1, x_2) = g_1(g_1(x_1, x_2), g_2(x_1, x_2))$. Use the Chain Rule (carefully!) to show that $\partial G/\partial x_1 = 1$ and $\partial G/\partial x_2 = 0$.

12. Let $f:\mathbf{R}^3 \to \mathbf{R}$ be differentiable and let $y, z:\mathbf{R} \to \mathbf{R}$ also be differentiable. Calculate $F'(x)$ where $F(x) = f(x, y(x), z(x))$. (If this is confusing, let $x(t) = t$ and find $(d/dt)(f(x(t), y(t), z(t)))$.)

13. The internal energy, u, of a gas can be expressed as a function of the pressure, volume and temperature of the gas, as $u = u(p, v, t)$. The pressure p may be expressed as a function of volume and temperature, $p = p(v, t)$. Thus the internal energy can be expressed in terms of v and t as $U(v, t) = u(p(v, t), v, t)$. Show that $\partial U/\partial t = (\partial u/\partial p) \cdot (\partial p/\partial t) + \partial u/\partial t$.

14. f is twice differentiable and $\forall x, y \in \mathbf{R}$ $f(x, y) = f(y, x)$. Show that

$$\frac{\partial f}{\partial x}(a, b) = \frac{\partial f}{\partial y}(b, a) \quad \text{and} \quad \frac{\partial^2 f}{\partial x^2}(a, b) = \frac{\partial^2 f}{\partial y^2}(b, a).$$

15. Let $A = \{\mathbf{x} \in \mathbf{R}^m : \|x\| \leq 1\}$ and $f:A \to \mathbf{R}$ be continuous and be differentiable at all points $\mathbf{a}$ satisfying $\|a\| < 1$. Prove that if f attains a maximum or minimum at an internal point of A (i.e. at a point $\mathbf{a}$ with $\|a\| < 1$), then $\partial f/\partial x_1 = \partial f/\partial x_2 = \cdots = \partial f/\partial x_m = 0$ there. Hence prove that if $f(\mathbf{x})$ is constant on the set $\{\mathbf{x}: \|x\| = 1\}$ then there is a point $\boldsymbol{\xi}$ with $\|\boldsymbol{\xi}\| < 1$ for which $\partial f/\partial x_i(\boldsymbol{\xi}) = 0$ $(i = 1, 2, \ldots, m)$.

16. Let $f:[a, b] \times [c, d] \to \mathbf{R}$ be continuous and define $F:[a, b] \times [c, d] \to \mathbf{R}$ by $F(x, y) = \int_c^y f(x, t) \, dt$. Prove that F is continuous.

17. Show that if $f(x, y)$ is continuous in the two variables separately and $\partial f/\partial y$ is jointly continuous, then f is jointly continuous.

18. Let $f:[a, b] \times [a, b] \to \mathbf{R}$ be continuous, and define $F, G:[a, b] \to \mathbf{R}$ by $F(z) = \int_a^z (\int_a^x f(x, y) \, dy) \, dx$, $G(z) = \int_a^z (\int_y^z f(x, y) \, dx) \, dy$. By showing that the result of the inner integration in both cases is continuous, and

differentiating under the integral sign, deduce that $F' = G'$ and hence that $F = G$. Sketch the sets over which integration is taken in the (x, y)-plane.

19. Given that

$$\int_0^{\pi/2} \frac{dx}{a^2 \sin^2 x + b^2 \cos^2 x} = \frac{\pi}{2ab},$$

show by differentiating under the integral sign that

$$\int_0^{\pi/2} \frac{\sin^2 x \, dx}{(a^2 \sin^2 x + b^2 \cos^2 x)^2} = \frac{\pi}{4a^3 b}$$

and find

$$\int_0^{\pi/2} \frac{dx}{(a^2 \sin^2 x + b^2 \cos^2 x)^2}.$$

20. Decide whether the functions given by the formulae below have a maximum, minimum or saddle point at $(0, 0)$: $xy, x^2 + y^2, x^2 + xy + y^2, x^4 + xy + y^2$.

21.* Let $A = \{x \in \mathbf{R}^2 : \|x\| < r\}$ and let $f_1, f_2 : A \to \mathbf{R}$ have continuous first order partial derivatives. By considering $F_1(t) = f_1(\mathbf{a} + t(\mathbf{b} - \mathbf{a}))$ show that if $f_1(\mathbf{a}) = f_1(\mathbf{b})$ where $\mathbf{a}$ and $\mathbf{b}$ are two distinct points of A then there is a point $\xi_1 \in A$ for which $(b_1 - a_1)(\partial f_1/\partial x_1)(\xi_1) + (b_2 - a_2)(\partial f_1/\partial x_2)(\xi_1) = 0$. If also $f_2(\mathbf{a}) = f_2(\mathbf{b})$ show that there is a second point $\xi_2 \in A$ for which $G(\xi_1, \xi_2) = 0$ where

$$G(\mathbf{x}, \mathbf{y}) = \frac{\partial f_1}{\partial x_1}(\mathbf{x}) \cdot \frac{\partial f_2}{\partial x_2}(\mathbf{y}) - \frac{\partial f_1}{\partial x_2}(\mathbf{x}) \frac{\partial f_2}{\partial x_1}(\mathbf{y}).$$

Deduce that if $G(\mathbf{0}, \mathbf{0}) \neq 0$, then for r sufficiently small, the function $\mathbf{x} \mapsto (f_1(\mathbf{x}), f_2(\mathbf{x}))$ is injective on A. (Notice that the condition $G(\mathbf{0}, \mathbf{0}) \neq 0$ is equivalent to the statement that the matrix

$$\begin{bmatrix} \dfrac{\partial f_1}{\partial x_1} & \dfrac{\partial f_1}{\partial x_2} \\ \dfrac{\partial f_2}{\partial x_1} & \dfrac{\partial f_2}{\partial x_2} \end{bmatrix}$$

of partial derivatives is non-singular at $\mathbf{0}$.)

Appendix A

SET THEORY

A **set** in mathematics is a well-defined collection of objects, well-defined in the sense that there is a criterion for membership such that, given an object, it is in principle possible to tell whether it belongs to the set or not. In our analysis we shall be concerned with sets of numbers or functions.

The notation $x \in A$ means that x **belongs to**, or **is an element of**, the set A. A set may be defined by listing its elements within curly brackets, e.g. $\{1, 2, 3\}$, the set whose elements are the numbers 1, 2 and 3, or by stating a property which defines the set: $\{x : x$ is a real number and $0 < x < 1\}$ denotes the set of all real numbers which lie strictly between 0 and 1. This is read as 'the set of all x such that x is a real number and $0 < x < 1$'.

There is no mystery about sets and we shall use them as a convenient notation, but notice two things: (i) it is possible to think of a set, the empty set, with no elements. This is denoted by $\varnothing$. (ii) an element either belongs to a set or it does not; it cannot belong more than once so that $\{1, 2, 3\}$ and $\{1, 2, 3, 3\}$ are notations for the same set.

A **subset** of a set is a set all of whose members belong to the original set. Thus A is a subset of B, written $A \subset B$, if and only if $x \in A \Rightarrow x \in B$.

If A and B are sets, we define the **union**, **intersection** and **difference**, denoted respectively by $A \cup B$, $A \cap B$ and $A \backslash B$ by

$$A \cup B = \{x : x \in A \quad \text{or} \quad x \in B\}$$
$$A \cap B = \{x : x \in A \quad \text{and} \quad x \in B\}$$
$$A \backslash B = \{x : x \in A \quad \text{and} \quad x \notin B\}.$$

($x \notin B$ means x is not an element of B.) Finally, the **Cartesian product**, $A_{\times} B$, of A and B is the set of ordered pairs (x, y) where $x \in A$ and $y \in B$.

$$A_{\times} B = \{(x, y) : x \in A \quad \text{and} \quad y \in B\}.$$

There are some specific sets for which we use a standard notation:

N, the set of all natural numbers; $\mathbf{N} = \{1, 2, 3, \ldots\}$.
Z, the set of all integers; $\mathbf{Z} = \{0, +1, -1, +2, -2, \ldots\}$.
Q, the set of all rational numbers; $\mathbf{Q} = \{p/q : p \in \mathbf{Z}, q \in \mathbf{N}\}$.
R, the set of all real numbers.

FUNCTIONS

Let A and B be two sets. A **function** f from A to B is a rule which associates with each element x of A a unique element $f(x)$ of B. (This prohibits, for example, the idea of defining a function by the rule $\pm \sqrt{x}$.) It is possible, and in courses on mathematical logic necessary, to make this more formal, but we shall not require to do so.

A is called the **domain** of f, and B the **codomain**. The notation $f:A \rightarrow B$ means f is a function with domain A and codomain B.

A function $f:A \rightarrow B$ is said to be **injective** if distinct points of A have distinct images under f, i.e. if $x \neq y \Rightarrow f(x) \neq f(y)$. (This is more readily checked in the form $f(x) = f(y) \Rightarrow x = y$.) f is said to be **bijective** if it is injective and every point of B is the image of some point of A. (The second condition states that for every y in B there is an x in A for which $f(x) = y$.)

Notice that there is a difference between the function f and its value at x, $f(x)$. We may specify a function by defining the rule, e.g. $f:\mathbf{R} \rightarrow \mathbf{R}$ where $f(x) = x^2$; this may be abbreviated $x \mapsto x^2$ if the domain is understood.

We denote by $f(X)$ and $f^{-1}(Y)$ the sets

$$f(X) = \{f(x) : x \in X\}$$
$$f^{-1}(Y) = \{x : f(x) \in Y\}$$

where X is a subset of the domain of f. If $f:A \rightarrow B$ then we call $f(A)$, the set of all points of B which are the image under f of some point of A, the **image** of f.

Appendix B

DECIMALS

At the end of Chapter 5 we claimed that every real number x can be expanded as a decimal, that is, that there are integers $b_N, b_{N-1}, \ldots, b_0, a_1, a_2, \ldots$, all belonging to the set $\{0, 1, \ldots, 9\}$, for which $x = \sum_{n=0}^{N} b_n 10^n + \sum_{n=1}^{\infty} a_n 10^{-n}$. We have to prove that such an expansion exists, to find whether this is unique and, if not, describe the various possible alternatives.

Firstly denote by $[x]$ the greatest integer which does not exceed x so that if x is real, $[x] \leq x < [x] + 1$. Then for $x \geq 0$ the integer $b_N b_{N-1} \ldots b_0$ in decimal notation, or $\sum_{n=0}^{N} b_n 10^n$, should equal $[x]$ and it is a simple exercise in induction to prove that every non-negative integer can be expressed in a unique way in the form $\sum_{n=0}^{N} b_n 10^n$ where each of the b_n is one of the digits $0, 1, \ldots, 9$. The remainder of the problem is to show that the fractional part of x is expressible as a decimal.

Let $0 \leq x < 1$. The first coefficient a_1 in the usual decimal expansion of x is the largest multiple of $1/10$ which does not exceed x, that is, $a_1/10 \leq x < (a_1 + 1)/10$. Therefore a_1 satisfies $a_1 \leq 10x < a_1 + 1$ and we see that $a_1 = [10x]$. With this definition we see that a_1 is an integer, $0 \leq a_1 < 10$ (since $0 \leq 10x < 10$) and $0 \leq x - a_1/10 < 1/10$. To determine a_2 we need to add the largest possible multiple of $1/100$ to $a_1/10$ for which the resulting sum is no greater than x. We therefore wish $a_1/10 + a_2/100 \leq x < a_1/10 + (a_2 + 1)/100$, whence, multiplying by 100, we see that the correct choice of a_2 is $[100(x - a_1/10)]$, and we obtain $0 \leq x - a_1 \cdot 10^{-1} - a_2 \cdot 10^{-2} < 10^{-2}$.

We now show by induction that for all $n \in \mathbf{N}$, a_n can be chosen from $0, 1, \ldots, 9$ so that, for all $N \in \mathbf{N}$,

$$0 \leq x - \sum_{n=1}^{N} a_n \cdot 10^{-n} < 10^{-N}. \tag{1}$$

By the work above, a_1 can be chosen satisfying (1). Suppose that for some N we have found $a_1, \ldots, a_N$ with the required properties. Then $0 \leq 10^{N+1}(x - \sum_{n=1}^{N} a_n \cdot 10^{-n}) < 10$ so if

$$a_{N+1} = \left[10^{N+1} \left(x - \sum_{n=1}^{N} a_n \cdot 10^{-n} \right) \right]$$

a_{N+1} is an integer between 0 and 9 and

$$a_{N+1} \leq 10^{N+1} \left(x - \sum_{n=1}^{N} a_n \cdot 10^{-n} \right) < a_{N+1} + 1$$

whence we see that (1) holds with $N + 1$ in place of N. Therefore we can find a_{N+1} so that $a_1, \ldots, a_{N+1}$ have the required properties. By induction, therefore, we can, for all $n \in \mathbf{N}$, find a_n so that a_n is an integer between 0 and 9 and, for all $N \in \mathbf{N}$, (1) holds.

Since $10^{-N} \to 0$ as $N \to \infty$, the chosen sequence has the property that $x = \sum_{n=1}^{\infty} a_n \cdot 10^{-n}$. Together with the expression of the integral part of x in decimal form, we have shown that every non-negative number has at least one decimal expansion. Call the expansion we have just constructed the *standard* expansion for x.

Suppose now that $b_1, \ldots, b_N$ are integers chosen from $0, 1, \ldots, 9$ and that for some N,

$$0 \le x - \sum_{n=1}^{N} b_n \cdot 10^{-n} < 10^{-N}. \tag{2}$$

Then, for $1 \le k \le N$,

$$0 \le x - \sum_{n=1}^{k} b_n \cdot 10^{-n} < \sum_{n=k+1}^{N} b_n \cdot 10^{-n} + 10^{-N}$$

$$\le \sum_{n=k+1}^{N} 9.10^{-n} + 10^{-N}$$

$$= (10^{-k} - 10^{-N}) + 10^{-N} = 10^{-k}.$$

In particular, $0 \le x - b_1 10^{-1} < 10^{-1}$ so $b_1 \le 10x < b_1 + 1$ whence $b_1 = [10x] = a_1$. By a similar argument, if $a_n = b_n$ for $n = 1, 2, \ldots, k$ then $b_{k+1} \le 10^{k+1}(x - \sum_{n=1}^{k} a_n \cdot 10^{-n}) < b_{k+1} + 1$, so $b_{k+1} = a_{k+1}$. It follows that $a_1, \ldots, a_N$ is the only sequence of N digits satisfying (2).

Now suppose that $x = \sum_{n=1}^{\infty} a_n \cdot 10^{-n}$, as before, but that x can also be written $x = \sum_{n=1}^{\infty} b_n \cdot 10^{-n}$ where each b_n is one of the digits $0, 1, \ldots, 9$. Then if $N \in \mathbf{N}$,

$$0 \le x - \sum_{n=1}^{N} b_n \cdot 10^{-n} = \sum_{n=N+1}^{\infty} b_n \cdot 10^{-n} \le \sum_{n=N+1}^{\infty} 9.10^{-n} = 10^{-N} \tag{3}$$

(since $\forall n \in \mathbf{N} \ 0 \le b_n \le 9$). If there is strict inequality on the right in (3), so that (2) is satisfied, then $a_n = b_n$ for $n = 1, 2, \ldots, N$ by our work above. There are, then, two possibilities: either the right-hand inequality in (3) is strict for all $N \in \mathbf{N}$ or there is at least one value of N for which $x - \sum_{n=1}^{N} b_n \cdot 10^{-n} = 10^{-N}$. In the first case, $a_n = b_n \ \forall n \in \mathbf{N}$ and we have the standard expansion. For the second case, let N be the smallest natural number such that $x - \sum_{n=1}^{N} b_n \cdot 10^{-n} = 10^{-N}$; there must be such an N since the set of all k for which $x - \sum_{n=1}^{k} b_n 10^{-n} = 10^k$ is a set of natural numbers which contains at least one element; hence it has a smallest element.

Let us look at the second case in more detail. $x - \sum_{n=1}^{N} b_n \cdot 10^{-n} = 10^{-N}$ and $10^{-N} = \sum_{n=N+1}^{\infty} 9.10^{-n} \ge \sum_{n=N+1}^{\infty} b_n 10^{-n}$, so $\forall n \ge N + 1 \ b_n = 9$, since if $b_n < 9$ for some n, the sums would differ. Now $b_N < 9$ for if $b_N = 9$ then $x - \sum_{n=1}^{N-1} b_n 10^{-n} = 9.10^{-N} + 10^{-N} = 10^{-N+1}$, contradicting the minimality of N (notice that since $x < 1$, it is not the case that $\forall n \ge 1 \ b_n = 9$). Therefore $x = \sum_{n=1}^{N} c_n 10^{-n}$ where $c_n = b_n \ (n = 1, \ldots, N - 1)$, $c_N = b_N + 1$ and since $b_N < 9$, c_N is one of $0, 1, \ldots, 9$ for $n = 1, 2, \ldots, N$. But then $0 = x - \sum_{n=1}^{N} c_n 10^{-n} < 10^{-N}$ so $c_1, \ldots, c_N$ satisfy condition (2) and therefore $c_n = a_n$ $(n = 1, \ldots, N)$. Moreover, since $0 = x - \sum_{n=1}^{N} a_n 10^{-n} = \sum_{n=N+1}^{\infty} a_n 10^{-n}$ and $a_n \geq 0 \; \forall n \geq N + 1$, we see that $\forall n \geq N + 1 \; a_n = 0$.

We have shown that the only way a number can have more than one decimal expansion is if it is of the form $m \cdot 10^{-N}$ for some integer m and $N \in \mathbf{N}$, in which case the two expansions are $\cdot a_1 a_2 \ldots a_N 000 \ldots$ and $\cdot a_1 a_2 \ldots a_{N-1} (a_N - 1) 999 \ldots$. It is easily checked that these two cases do occur.

A decimal is said to repeat if there exist integers N, k such that $\forall n \geq N$ $a_{n+k} = a_n$. Every repeating decimal represents a rational number, for if $x = \sum_{n=1}^{\infty} a_n \cdot 10^{-n}$ and $\forall n \geq N$ $a_{n+k} = a_n$ then $x = \sum_{n=1}^{N-1} a_n \cdot 10^{-n} + \sum_{n=N}^{\infty} a_n 10^{-n}$. The first of these sums is clearly a rational number. Also if we let $y = \sum_{n=N}^{\infty} a_n \cdot 10^{-n}$ then $y = \lim_{m \to \infty} s_m = \sum_{n=N}^{m} a_n \cdot 10^{-n}$. Now if $m = N + jk - 1$,

$$s_m = \sum_{m=N}^{N+jk-1} a_n 10^{-n} = \left(\sum_{n=N}^{N+k-1} a_n \cdot 10^{-n} \right)(1 + 10^{-k} + \cdots + 10^{-(j-1)k})$$

$$= \alpha \cdot (1 - 10^{-jk})(1 - 10^{-k})^{-1}$$

where $\alpha = \sum_{n=N}^{N+k-1} a_n \cdot 10^{-n}$. $\lim_{j \to \infty} s_{N+jk-1} = \alpha(1 - 10^{-k})^{-1}$. Now $\forall \varepsilon > 0 \; \exists M$ s.t. $\forall m \geq M \; |s_m - y| < \varepsilon/2$, therefore (choosing m of the form $N + jk - 1$ for $j \geq J$ and $m \geq M$, where $\alpha \cdot 10^{-Jk}(1 - 10^{-k})^{-1} < \varepsilon/2$, $|y - \alpha(1 - 10^{-k})^{-1}| \leq |y - s_m| + |s_m - \alpha(1 - 10^{-k})^{-1}| < \varepsilon$. Since ε was arbitrary, $y = \alpha(1 - 10^{-k})^{-1}$ so y, and hence x, is rational.

Interestingly, the converse to this is true; every rational number has a repeating decimal. (Notice that $.50000 \ldots$ repeats in this sense.) Suppose p and q are integers and $0 < p/q < 1$. Then, using the technique learned in school, we would calculate the decimal expansion for p/q by long division. Let $10p = a_1 q + r_1$ where a_1, r_1 are integers and $0 \leq r_1 < q$. Then let $10r_1 = a_2 q + r_2$, and so on, so that at the nth stage, $10r_{n-1} = a_n q + r_n$ with $0 \leq r_n < q$. It is easy to check that

$$0 \leq p/q - \sum_{n=1}^{N} a_n \cdot 10^{-n} = (r_N/q) \cdot 10^{-N} < 10^{-N}$$

so that this process does indeed yield the correct expansion. Moreover, there are at most q possible remainders, so after $q + 1$ steps the same remainder must have occurred more than once and the pattern will then be repeated thereafter. The remaining details are left as a problem.

References

Apostol, T.M. (1974), Mathematical Analysis, Addison Wesley (Chapter 11).

Brown, A.L. and Page, A. (1970), *Elements of Functional Analysis*, Van Nostrand Rheinhold.

Dunning-Davies, J. (1982), *Mathematical Methods for Mathematicians, Physical Scientists and Engineers*, Ellis Horwood.

May, R.M. (1976), 'Simple mathematical models with very complicated dynamics', *Nature*, **261**, 10 June 1976, 459–467.

Taylor, A.E. (1958), *Introduction to Functional Analysis*, Wiley.

Hints and Solutions to Selected Problems

Chapter 2.
1. (i) $x = 0$ (ii) $x = \sqrt{3}/2$ (iii) $x = \pi/4 + k\pi$ for k an integer.
2. $\Rightarrow; \Leftrightarrow; \Leftarrow; \Leftrightarrow; \Rightarrow; \Leftarrow$.
3. (i) $x^2 + xy + y^2 = (x + \frac{1}{2}y)^2 + 3y^2/4$
   (ii) $\Leftarrow$ and $\Leftrightarrow$ respectively. (Use $x^3 - y^3 = (x - y)(x^2 + xy + y^2)$.)

Chapter 3.
1. Use property A12 and deal with $x = 0$ case separately.
2. Use Q1 for $\Rightarrow$. For $\Leftarrow$, notice $a^2 - x^2 = (a - x)(a + x)$.
3. $0 < x < y \Rightarrow 0 < 1 + x < 1 + y \Rightarrow 0 < 1/(1 + y) < 1/(1 + x)$.
5. $x^3 \pm 3x + 2 = (x \pm 3/2)^2 - 1/4$.
6. $x + 1/x - 10/3 = (x^2 - 10x/3 + 1)/x = \{(x - 5/3)^2 - 16/9\}/x$. Show that for $x > 3$ this is positive.
7, 8. ordinary induction. For end of Q8, notice first part proves result for $n \geq 2$.
9. $a_{n+1} = (a_n - 1)^2 + 1$.
10. (ii) Put $x = k + 1$, $y = k$ in (i). For (iii), use induction and part (ii).
11. If n is not divisible by 3, $n = 3m + 1$ or $3m + 2$ for an integer m.
12. Assume p, q have no common factor and $p^3 = 2q^3$ and prove a contradiction.
13, 14. By contradiction. Also, n^2 divisible by $6 \Rightarrow n^2$ div. by 2 and 3.
15. If $a\sqrt{2} + b\sqrt{3}$ is rational, so is $(a\sqrt{2} + b\sqrt{3})^2$. Obtain contradiction.
16. If p, q both odd, the sum of three odd numbers is 0; impossible. For the last part, let $x = p/q$ where p, q have no common factor and use contradiction.
17. $10^{m+1} - 1 = 10(10^m - 1) + 9$ gives induction step to prove 9 divides $10^n - 1$.
18. Prove $10^m + (-1)^{m-1}$ is divisible by 11 for all $m \in \mathbf{N}$.
19. (i) Let $m * 1 = 1 * m$. Then $1 * (m + 1) = (1 * m) + 1 = (m * 1) + 1$
$$= (m + 1) + 1 = (m + 1) * 1$$
all by definition of $*$ except the $m * 1 = 1 * m$ step. This is the induction step.
(ii) Let $P(n)$ be '$\forall m \in \mathbf{N} \; m * (n + 1) = (m + 1) * n$'. $P(1)$ is true (use definition of $*$). Then
$$P(n) \Rightarrow \forall m \; m * ((n + 1) + 1) = (m * (n + 1)) + 1$$
$$= ((m + 1) * n) + 1 = (m + 1) * (n + 1).$$

163

(iii) Let $P(n)$ be '$\forall m \in \mathbb{N}$ $m*n = n*m$'.

$P(n) \Rightarrow \forall m \in \mathbb{N}$ $m*(n+1) = (m+1)*n = n*(m+1) = (n+1)*m$ (by (ii), $P(n)$ and (ii) respectively).

20. $(3/2)^n > n/2$.
21. $2^n > (5/4)^{3n}$ and $(5/4)^n > n/4$ whence $(5/4)^{3n} > n^3/64$. Hence $n^2/2^n < 64/n$.
22. $|(x+y)+z| \le |x+y| + |z|$.
25. Notice $|\,|x| - |y|\,| = |x| - |y|$ or $|y| - |x|$.
28. $|2+x| \ge |2 - |-x|\,| \ge 2 - |x|$, so $|x| \le 1 \Rightarrow |2+x| \ge 1$. For the last part, put over denominator $|x+2|$ and notice that if $\delta_2 \le 1$ then $1/|x+2| \le 1$.
29. $\delta \le 1/2 \Rightarrow |x| = |1 - (1-x)| \ge 1/2 \Rightarrow 1/|x| \le 2$ in (i).
30. Prove by contradiction.
31. $x = -2$, $y = -1$; $x = -a-1$, $y = -a$; $x = 2$, $y = -2$.

Chapter 4.

1. (i) $|a_n - 1| = 2/(n+1)$. Let $\varepsilon > 0$. $\exists N$
   (e.g. $N > 2/\varepsilon$) s.t. $\forall n \ge N$ $|a_n - 1| < \varepsilon$.
   (ii) $|a_n - 1| \le \min(1/n, 2/n^2) \le 2/n$. (iii) $|a_n - 0| < 1/n$.
2. $\forall n \ge N$ $|b_n - L| < \varepsilon \Leftrightarrow \forall n \ge N+1$ $|a_n - L| < \varepsilon$.
3. Find n s.t. $\forall n \ge N$ $a_n > L - \varepsilon$ and $c_n < L + \varepsilon$.
4. $1/2$, 0, 0, no limit.
7. If (a_n) is not bounded above, $\forall R$ $\exists N$ s.t. $a_N > R$.
9. $a_{n+1} = 1 + (a_n - 1)^2/2$. Show $\alpha \in [1, 3) \Rightarrow \forall n$ $a_n \in [1, 3)$, and consider $a_{n+1} - a_n = (a_n - 1)(a_n - 3)/2$.
10. If α belongs to one of $[0, 1), (1, \infty), [-1, 0]$ then all the terms (a_n) belong to the same interval. Use $a_{n+1} - a_n = a_n(a_n - 1)/2$ thereafter.
11. Increasing if $\alpha \in (0, 1]$, decreasing if $\alpha \in [1, \infty)$. $\alpha < -1 \Rightarrow a_2 > 2$.
12. $1 \le a_n \le (1 + \sqrt{5})/2 \Rightarrow 1 \le a_{n+1} \le a_n$ ($\le (1 + \sqrt{5})/2$).
15. $a_n = (-1)^n n$.
17. If $n \ge 2N - 1$, then $b_n = a_m$ where $m \ge N$. If $n \ge N(N-1)/2 + 1$ then $c_n = a_m$ for some $m \ge N$.
18. For first part, $a_n = 0$, $b_n = 1/n$. In second, $a + A \le b$ by Th. 4.2.
19. (ii) If $a \ne 0$, $|\,|a_n| - |a|\,| = |a_n^2 - a^2|/(|a_n| + |a|) \le |a_n^2 - a^2|/|a|$.
   (iii) $a_n = (-1)^n$.
   (iv) $a_n - a = (a_n^3 - a^3)/(a_n^2 + aa_n + a^2)$ and $a_n^2 + aa_n + a^2 = (a_n + a/2)^2 + 3a^2/4 \ge 3a^2/4$.
   (v) Easy if $a = 0$, so let $a \ne 0$. By (ii), $|a_n| \to |a|$. If $|a_n| - |a|, |a_{n+1}| - |a|$ and $|a_{n+1} - a_n|$ are all less than $|a|/2$, a_n and a_{n+1} have same sign.

Chapter 5.

2. $0 \le a_n \le (3/4)^n$.
3. D, D, C, C. (D = divergent, C = convergent.)
4. Choose N s.t. $\forall n \ge N$ $0 \le a_n < 1$ ($a_n \to 0$); then $\forall n \ge N$ $0 \le a_n^2 < a_n$. Let $b_n = 1/n$.
6. All converge except $\sum (2n)!/(n!)^2$.
7. All positive α by alternating series test.
8. Convergent for all x except ± 1; use ratio test (except $x = \pm 1$).

9. Radii: 1, 1/2, 1, 1, 1, $\sqrt{2}$, 2, 1, 0, ∞, 1/4, 4/27, ∞ where ∞ denotes that the series converges for all real x.

10. Values of x, with $|x| =$ radius, for which series converges: ± 1, neither, ± 1, -1, neither, neither, -2, neither, 0, inapplicable.

13. From the limit $\exists N$ s.t. $\forall n \geq N$ $L - \varepsilon < |a_{n+1}/a_n| < L + \varepsilon$. Write $|a_n| = |a_n/a_{n-1}| \cdot |a_{n-1}/a_{n-2}| \cdots |a_{N+1}/a_N| \cdot |a_N|$.

16. By contradiction.

17. $\sum c_n$ and $\sum b_n$ cannot both converge (since $\sum |a_n|$ diverges); by Q16, neither converges since $\sum a_n$ is not divergent.

18. Let $t_n = b_1 + \cdots + b_n$ and $s_n = a_1 + \cdots + a_n$. $t_n = s_{2n}$. Try modifying $\sum a_n$ by putting in more zeros.

19. Find $\sum_{n=1}^{k} c_n$.

20. $q! |p/q - s_{q+1}| \leq q(q+2)/(q+1)^2 < 1$. The only integer whose modulus is less than 1 is 0, so $p/q = s_{q+1}$.

21. $\binom{n}{k} \dfrac{1}{n^k} = \dfrac{n(n-1) \cdots (n-k+1)}{n^k} \dfrac{1}{k!} = 1 \cdot (1 - 1/n) \ldots (1 - (k-1)/n) \dfrac{1}{k!} \leq \dfrac{1}{k!}$.

Then use Binomial Theorem.

Chapter 6.

1. $[1, \infty)$, $[\frac{1}{2}, \infty)$, $[4, \infty)$, $[4, \infty)$, $[1, \infty)$, $[\frac{1}{2}, \infty)$.

2. 2 is *an* upper bound for A; sup A is the smallest upper bound.

4. Prove rq irrational by contradiction; $1/q \in \mathbf{Q}$. For the last part, prove there is a non-zero rational q with $x/\sqrt{2} < q < y/\sqrt{2}$.

6. Suppose we have defined $b_n \in \mathbf{Q}$ with $\sqrt{2} - 1/n < b_n < \sqrt{2}$. Choose $b_{n+1} \in \mathbf{Q}$ with $\max(b_n, \sqrt{2} - 1/(n+1)) < b_{n+1} < \sqrt{2}$. Use induction to show (b_n) has the required properties.

7. Prove that $x \in B \Rightarrow x \leq 1 + \sup A$ also that $b < 1 + \sup A \Rightarrow b - 1 < \sup A$ hence $\exists x \in A$ with $x > b - 1$ so $\exists y \in B$ with $y > b$. Similar idea for C; show $x \in C \Rightarrow x \leq 2$ sup $A - 1$ and that $c < 2$ sup $A - 1 \Rightarrow \exists x \in C$ with $x > c$. (To do last piece, relate a number involving c to sup A.)

8. Show $1/\sup A$ is a lower bound for B, and either use the second condition in Lemma 6.8 or obtain inequalities for inf B by relating sup A to $1/\inf B$.

9. Again Lemmas 6.3 and 6.8. Notice that $c > \inf A + \inf B \Rightarrow c - \inf A > \inf B \Rightarrow \exists x \in B$ s.t. $c - \inf A > x \Rightarrow c - x > \inf A$.

10. Let $c < \sup A \cdot \sup B$; then $c/\sup A < \sup B$ since sup $A > 0$ (why?). This gives the start of the second condition in Lemma 6.3 for sup C. The case for inf is similar if inf $A > 0$ and inf $B > 0$. The only remaining case, inf A inf $B = 0$, is easy.

Chapter 7.

1. $|x - a| < \delta \Rightarrow |x^3 - a^3| \leq |x - a|(|x|^2 + |ax| + |a|^2) < \delta((|a| + \delta)^2 + |a|(|a| + \delta)| + |a|^2)$ so if $\varepsilon > 0$, $|x - a| < \delta = \min(1, \varepsilon/(3|a|^2 + 3|a| + 1)) \Rightarrow |x^3 - a^3| < \varepsilon$.

3. $|x| \leq 1 \Rightarrow |x^2| \leq |x|$, so $|f(x) - 1| \leq |x|$.

5. $|f(x) - f(0)| \leq K|x|$.

6. Let $\varepsilon = f(a)$.

7. $|x| \geq 1 \Rightarrow |a_2/x + a_1/x^2 + a_0/x^3| \leq (|a_1| + |a_2| + |a_3|)/|x| < \varepsilon$ if $x > \max(1, (|a_1| + |a_2| + |a_3|)/\varepsilon)$.

9. Let $c \in \mathbf{R} \backslash \mathbf{Q}$. Let $\varepsilon > 0$. $\exists \delta > 0$ s.t. $|x - c| < \delta \Rightarrow |g(x) - g(c)| < \varepsilon$. Observe that there is at least one rational x satisfying $|x - c| < \delta$.

11. $\exists X$ s.t. $\forall x \geq X$ $L - 1 < f(x) < L + 1$ (because $f(x) \to L$ as $x \to \infty$). $\{f(x) : 0 \leq x \leq X\}$ is bounded because f is continuous.

12. Let $a \in \mathbf{R}$. If $g(a) > f(a)$, then $\exists \delta > 0$ s.t. $|x - a| < \delta \Rightarrow f(x) < g(a)$ so $|x - a| < \delta \Rightarrow g(x) = g(a)$. If $g(a) = f(a)$ and $\varepsilon > 0$, $\exists \delta > 0$ s.t. $|x - a| < \delta \Rightarrow g(a) - \varepsilon < f(x) < g(a) + \varepsilon$ whence $|x - a| < \delta \Rightarrow g(a) - \varepsilon < g(x) \leq g(a) + \varepsilon$.

13. Let $y \in \mathbf{R}$. Choose x_0 and x_1 s.t. $h(x_0) < y < h(x_1)$ (possible since h is unbounded above and below).

Chapter 8.

1. $f'(x) = 1$ if $x > 0$ and $f'(x) = -1$ if $x < 0$.

3. $2f'(2x)$, $2xf'(x^2)$, $f'(g(x) \cdot h(x))(g'(x)h(x) + g(x)h'(x))$, $f'(g(h(x))) \cdot g'(h(x)) \cdot h'(x)$.

5. Use Mean Value Theorem.

6. Show $g' = 0$.

8. Use Rolle on $[a, b]$ and $[b, c]$, then apply it to f'.

10. (ii) Apply part (i) to f' to show $\exists \delta > 0$ s.t. $x \in [b, b + \delta) \Rightarrow f'(x) > 0$, then use MVT to show f strictly increasing on $[b, b + \delta)$.

12. $-1, -2, 1, 1/2$.

13. f is strictly decreasing on $(0, \infty)$, hence $3/2 \leq x \leq 2 \Rightarrow f(2) \leq f(x) \leq f(3/2)$. Since $a_2 = 2$, $\forall n \geq 2$ $a_n \in [3/2, 2]$. Let $a = (1 + \sqrt{5})/2$, so $a = f(a)$; use MVT on $f(a_n) - f(a)$, noting that $|f'(\xi_n)| \leq 4/9$.

14. Apply method of Q13.

16. $|(f(x) - f(0))/x| = |x \sin(1/x)| \leq |x|$.

17. If f is not injective, $\exists x \neq y$ s.t. $f(x) = f(y)$; apply Rolle. Then use Lemma 7.7. Finally use Q9 to show $f'(c) < 0 < f'(d)$ implies f is neither strictly increasing nor strictly decreasing.

Chapter 9.

1. Radii: $1, 2, 1/4, 0, \infty, 1, 1/\sqrt{2}$.

3. From the limit, $\exists X$ s.t. $\forall x \geq X$, $(\alpha \log x)/x < \frac{1}{2}$ so $\alpha \log x < x/2$ and $\alpha \log x - x < -x/2$. Take exponentials for $x^\alpha e^{-x}$.

4. $x^x = \exp(x \log x)$.

7. (ii) Let $a_n = (-1)^n x^{2n+1}/(2n + 1)!$ $b_n = (-1)^n y^{2n}/(2n)!$ in the Cauchy Product theorem to obtain $\sin x \cos y$, and a similar use for $\cos x \sin y$.
   (iii) Differentiate w.r.t. x.

9. Use Theorems 9.10 and 9.11.

10. Let $f(x) = x - \log(1 + x)$, $g(x) = x - \log(1 + x) - x^2/2$. $f(0) = g(0) = 0$, and $x > 0 \Rightarrow f'(x) > 0$, $g'(x) < 0$ so $\forall x \geq 0$ $f(x) \geq 0$, $g(x) \leq 0$. Calculate $a_{n+1} - a_n$ and notice $a_{n+1} - a_1 = \sum_{i=1}^{n} (a_{i+1} - a_i)$.

11. $y = \sum_{n=1}^{\infty} (n - 1)! \, x^n$, radius 0.

12. $y^m e^{-y} \to 0$ as $y \to \infty$ so $x^{-m} e^{-1/x} \to 0$ as $x \to 0+$. Use this to show $(f^{(n)}(x) - f^{(n)}(0))/x \to 0$ as $x \to 0+$.

Chapter 10.

3. Show that if D is a dissection, $s(f; D) \geq \alpha \cdot 2\delta$.

5. Put $\phi = f$ and apply Q4 to $|f^2|$.

6. Apply IVT to $\int_a^x f - \int_x^b f$.

9. Use the integral to estimate $\log(1 + 1/n)$, and use the inequality to estimate $a_{n+1} - a_n$. Then induction.

10. $\int_1^n \log t \, dt = n \log n - n + 1$. By Trapezium Rule, $\int_1^n \log t \, dt = \sum_{i=1}^n \log i - (\log n)/2 + \sum_{i=1}^{n-1} \varepsilon_i$. $\sum \varepsilon_i$ converges.

11. If $a_n \to L \neq 0$ as $n \to \infty$, then $a_n^2/a_{2n} \to L$ as $n \to \infty$. Use this with $a_n = n!(\sqrt{n}(n/e)^n)^{-1}$.

12. Calculate; compare with e^{-ax}; integrate by parts and compare with $(\cos x)/x^{\alpha+1}$; calculate; calculate; compare with $x^{-1/2}$ on $(0, 1/2)$ and with $(1 - x)^{-1/2}$ on $(\frac{1}{2}, 1)$; calculate.

14. Integrate $(\cos x)/\sqrt{x}$ by parts and consider $\int_1^\infty (\sin x)/x^{3/2} \, dx$; for $\int_0^1$ compare with $1/\sqrt{x}$.

16. If $f(x) \to L$ as $x \to \infty$ (through real values), then $f(2n\pi) \to L$ as $n \to \infty$ (through integer values). Express $\int_{(2n-1)\pi}^{2n\pi}$ in terms of $\int_{(2n-2)\pi}^{(2n-1)\pi}$.

17. $\int_0^\infty (x^2 - 1)^{-1} \, dx$ does not exist (behaviour near $x = 1$). The second integral exists, by comparison with $1/x^2$ on $[2, \infty)$, with $1/\sqrt{|x-1|}$ on $(0, 1)$ and $(1, 2)$. The third can be calculated (by parts). The last does not exist; if $\alpha \leq -1$ the integral diverges on $(0, 1)$ and if $\alpha \geq -1$ it diverges on $(1, \infty)$.

19. Let M be an upper bound for $|f|$. Choose x so that $0 < x - a < \varepsilon/(4M)$, and let D_0 be a dissection of $[x, b]$ with $S(D_0) - s(D_0) < \varepsilon/2$. Let D be the points of D_0 together with a.

21. (i) is true because $f \geq 0$ and $0 \leq \int_c^d f = -\int_a^c f - \int_d^b f \leq 0$.
    (ii) Since $\inf\{S(D): \text{all } D\} = 0$ and $b - a > 0$, $\exists D_1$ with $S(D_1) \leq b - a$. Thus on at least one subinterval of D_1, $M_i = \sup\{f(x): x_{i-1} \leq x \leq x_i\} \leq 1$.
    (iii) Let D_{n+1} be a dissection of $[a_n, b_n]$ with $S(D_{n+1}) \leq (b_n - a_n)/(n + 1)$. The sup of f on at least one subinterval of D_{n+1} must be $\leq 1/(n + 1)$.

Chapter 11.

1. The first and third are continuous at $(0, 0)$, the others discontinuous.

3. Let $h(z) = (\sin z)/z$ $(z \neq 0)$, $h(0) = 1$; h is continuous, and so is $h(x + y)$ as a function $\mathbf{R}^2 \to \mathbf{R}$. For the second, use MVT to notice that $\sin x - \sin y = (x - y)\cos \xi$ for some ξ between x and y.

5. The first two are closed, the third not. $(-1 + 1/n, 0) \to (-1, 0)$ as $n \to \infty$.

6. Choose N s.t. $\forall m$, $n \geq N$ $|x_m - x_n| < 1$. Then $\forall m \geq N$ $|x_m| < 1 + |x_N|$ (setting $n = N$); hence show (x_m) bounded.

7. $F(t) = f(\mathbf{b} + t(\mathbf{c} - \mathbf{b}))$ for $0 \leq t \leq 1$.

8. In all cases, $\partial f/\partial x = \partial f/\partial y = 0$ at $(0, 0)$. The last two are differentiable.

11. $G(x_1, x_2) = f(g_1(x_1, x_2), g_2(x_1, x_2))$ where $f(y_1, y_2) = g_1(y_1, y_2)$ so $\partial G/\partial x_1 = (\partial f/\partial y_1) \cdot (\partial g_1/\partial x_1) + (\partial f/\partial y_2)(\partial g_2/\partial x_1)$; now substitute.

14. Let $g_1(x, y) = x$ and $g_2(x, y) = y$. Then let $F(x, y) = f(g_2(x, y), g_1(x, y))$; use this, noting that $F = f$.

16. $F(x, y) - F(x_0, y_0) = \int_{y_0}^y f(x, t) \, dt + \int_c^{y_0} (f(x, t) - f(x_0, t)) \, dt$; use uniform continuity and boundedness of f.

19. Differentiate under the integral sign w.r.t. a and b.
21. The first part is Rolle's Theorem applied to F_1; apply it also to F_2 where $F_2(t) = f_2(\mathbf{a} + t(\mathbf{b} - \mathbf{a}))$. This gives two simultaneous equations for the 'unknowns' $b_1 - a_1$, $b_2 - a_2$ with a non-trivial solution; the condition for $G(\xi_1, \xi_2)$ arises by considering the coefficients in these equations. Then use continuity of G.

Notation Index

A stroke through a symbol denotes 'not', so $a \notin A$ means that a does not belong to A. The same applies to $\nless, \ngtr, \nleq, \ngeq, a_n \nrightarrow a$, etc.

Subject Index

absolute convergence, 54, 57, 130, 132
absolute value (modulus), 27
alternating series theorem, 55
Archimedean property, 36, 66
arithmetic, 16
attainment (of sup, inf), 78, 142
axioms (for **R**), 16, 63–64

Bernoulli's Inequality, 26
Bessel's equation, 110
bijective function, 158
binomial theorem, 109
'blunt' inequality ($\leq$ or $\geq$)
Bolzano–Weierstrass Theorem, 141
bounded (and bounded above/below),
    42, 64, 65, 68, 77, 141
boundedness of continuous function,
    77, 142
bounds (upper/lower), 42, 64, 77
brackets, adding or removing from
    series, 57

calculus, fundamental theorem of, 124
Cauchy Product Theorem, 104
Cauchy's condensation test, 52, 131
Cauchy's mean value theorem, 89
Cauchy sequence, 154
Chain Rule, 86, 145
chaos, 96
closed interval, 64, 69
closed set, 139, 142
codomain of function, 158
coefficients of power series, 102, 110
comparison test (for series), 51–54
comparison test (for integrals), 130, 132
complete induction, 22
completing the square, 15, 20
composition of functions, 75, 85
condensation test, 52, 131
continuous function, 74ff, 119, 120, 137,
    139, 141, 142
continuity

axiom of, 64
joint, 139
of function, 74ff, 119, 120, 137, 139,
    141, 142
of inverse function, 80
on left/right, 75
relation to integrals, 120, 142
relation to sequences, 92, 139
separate, 139
uniform, 120, 141
contradiction, proof by, 23ff
convergence
    of sequence (= tending to a limit), *see*
        limit
    of series
        definition, 49
        tests, 51ff, 130
        radius of, 59, 100
cosine function, 112

decimal notation, 32, 59, 159
decimals, 59, 159ff
decreasing function, 80, 118
decreasing sequence, 43, 140
Dedekind's axiom, 64
derivative, 84ff, 143
difference of sets, 157
differentiable/differentiability, 84ff, 143ff
differential equation, 110, 151
differentiation under the integral sign,
    125, 148
dissection of interval, 116
distance in $\mathbf{R}^n$, 137
divergence of series, 49
domain of function, 158

e, 60, 61, 107
exponential function, 104, 107ff

for all, 21, 42
Fibonacci numbers, 23

partial derivative, 143ff
parts, integration by, 126
power of a real number, 109
power series, 58, 100ff
prime number, 22, 25
product of power series, 104

radius of convergence, 59, 100
ratio test, 54, 60 (Q13)
rational number, 24, 67, 161
real numbers, axioms for, 16, 63–64
rearranging series, 57
recurrence, 92
remainder in Taylor's theorem, 90–91, 111, 126
reversing order of integration, 149
reversing order of limits, 46, 148ff
Riemann–Lebesgue lemma, 135 (Q20)
right-hand limit, 73
right continuity, 75
Rolle's theorem, 87
root test, 54, 60 (Q13)
roots (nth), 79, 81
roots, using Newton's method to find, 96

saddle point, 151
sandwich rule', 47
separate continuity, 139
sequence, 34, 139
   bounded increasing, 43, 65
   boundedness of, 42, 141
series (see also power series), 49ff

adding or removing brackets, 57
rearrangement of, 57
set, 157
   closed, 139–142
sets of natural numbers, 25
Simpson's rule, 129
sine function, 112
smallest upper bound (= supremum), 65
stationary point, 87, 91, 150
strictly increasing/decreasing, 43, 80
subsequence, 140–141
substitution, integration by, 126
sum of series, 49
supremum, 65
supremum, attained by continuous function, 78, 142

Taylor's theorem, 90, 111, 126, 150
tests for convergence of series, 51ff, 130
tests for convergence of integrals, 130ff
trapezium rule, 128
triangle inequality, 28, 32 (Q25)
turning point, 87, 91, 150

union of sets, 157
uniform continuity, 120, 141
upper bound, 42, 64, 65, 68
upper bound, existence of least, 64
upper sum, 116

Weierstrass, 34, 141

Statistics and Operational Research
Editor: B. W. CONOLLY, Professor of Operational Research, Queen Mary College, University of London

Beaumont, G.P.	**Introductory Applied Probability**
Beaumont, G.P.	**Probability and Random Variables**
Conolly, B.W.	**Techniques in Operational Research: Vol. 1, Queueing Systems**
Conolly, B.W.	**Techniques in Operational Research: Vol. 2, Models, Search, Randomization**
Conolly, B.W.	**Lecture Notes in Queueing Systems**
French, S.	**Sequencing and Scheduling: Mathematics of the Job Shop**
French, S.	**Decision Theory: An Introduction to the Mathematics of Rationality**
Griffiths, P. & Hill, I.D.	**Applied Statistics Algorithms**
Hartley, R.	**Linear and Non-linear Programming**
Jolliffe, F.R.	**Survey Design and Analysis**
Jones, A.J.	**Game Theory**
Kemp, K.W.	**Dice, Data and Decisions: Introductory Statistics**
Oliveira-Pinto, F.	**Simulation Concepts in Mathematical Modelling**
Oliveira-Pinto, F. & Conolly, B.W.	**Applicable Mathematics of Non-physical Phenomena**
Ratschek, J. & Rokne, J.	**Computer Methods for the Range of Functions**
Ratschek, J. & Rokne, J.	**New Computer Methods for Global Optimisation**
Schendel, U.	**Introduction to Numerical Methods for Parallel Computers**
Stoodley, K.D.C.	**Applied and Computational Statistics: A First Course**
Stoodley, K.D.C., Lewis, T. & Stainton, C.L.S.	**Applied Statistical Techniques**
Thomas, L.C.	**Games, Theory and Applications**
Whitehead, J.R.	**The Design and Analysis of Sequential Clinical Trials**